Building VMware NSX® Powered Clouds and Data Centers for Small and Medium Businesses

Shahzad Ali, VMware

Foreword by Chris McCain, Director of Product Management, NSBU

vmware· PRESS

VMWARE PRESS

Program Managers

Katie Holms
Shinie Shaw

Technical Writer

Rob Greanias

Reviewers

Brant Scalan
Josh Newton
Nauman Mustafa
Nimish Desai
Gustavo Santana

**Designer and
Production Manager**

Michaela Loeffler
Sappington

**VMware, Inc. 3401 Hillview Avenue Palo Alto CA 94304 USA
Tel 877-486-9273 Fax 650-427-5001 www.vmware.com.**

Table of Contents

List of Figures

List of Tables

About the Author

Shahzad Ali
Lead Solutions Architect
VMware Inc.

Shahzad Ali leads the data center architect and solutions team in VMware. Shahzad's passion and experience is to take products and ideas from concept to market success. He is obsessed with helping customers solve their business challenges through innovative ideas and cutting-edge technologies.

Shahzad is a sought-after public speaker and has presented at conferences across the globe. He has been an academic guest lecturer at the graduate level on various topics related to cloud computing and data center designs. His previous published work includes authoring various networking, collaboration, and data center design and implementation guides, including *Designing NSX for Small Data Centers*.

Shahzad holds Bachelor's degree in Computer Science, Master's Degree in Electrical Engineering with networking major, Diploma in Product Management and Marketing. He is distinguished as a thought leader and influencer among his peers.

You can contact Shahzad on Twitter @virtualShahzad.

Acknowledgements

First and foremost, I would like to thank Almighty God for everything. I would like to thank my dad, my daughters Nabeeha and Saamiyah for motivating and encouraging me. My lovely wife, Sajida, for supporting me – not only during the writing of this book but with all endeavors throughout my technical career. None of this was possible without your support. I love you all. A strong shout out to the reviewers, with special thanks to Brant Scalan who provided more than just a technical review by suggesting content modification where necessary to make the book more comprehensible and fluent.

Thank you also to our VMware NSX customers for providing me ideas and platform to validate some of the concepts in their production network, allowing me to bring the real-world knowledge to our larger audience.

Finally, a call out to entire NSBU customer success solutions architects and management team, NSBU product marketing, product management, and tech-writers. Special thanks to Katie Holms and Shinie Shaw. Thanks for the support in getting this project started and the momentum going. This would not have happened without you and your extended team.

Shahzad Ali

Shahzad Ali, Lead Solutions Architect, VMware Inc.

Book Format

This book begins with a discussion around the main VMware NSX® for vSphere® use cases. Readers are introduced to VMware NSX and its management, control, and data plane components.

The discussion then moves on to relevant NSX deployment models used in small data centers, offering insight into central aspects of design. After the design and deployment model discussion, the book focuses on implementation aspects of individual NSX components.

Day 2 operations and troubleshooting topics are extremely important for the success of any deployment. The book examines these concepts, then closes with a discussion on strategies to grow NSX beyond a small data center footprint.

Book Objectives

The main objective of writing this book is to provide NSX design and deployment guidance for small and medium businesses (SMBs) with a smaller VMware ESXi™ footprint. Regardless of the size of a data center, VMware NSX enables features and use cases that provide business-impacting benefits of agility, reduced CAPEX/OPEX, flexibility, enhanced operations, and monitoring.

This book also discusses design deviations that are permitted for a SMB data center when compared to the reference design and deployment model for large enterprise data centers.

Intended Audience

This book is targeted toward virtualization, network, and security administrators, along with architects interested in designing and deploying the NSX network virtualization solution in a vSphere environment. It focuses on design and implementation aspects of NSX for vSphere (NSX-V) – the flagship product for vSphere customers using VMware ESXi as their sole data center hypervisor.

Foreword

The most important data center is yours; the one you are responsible for. Your livelihood depends on the key design qualities of that data center; availability, manageability, performance, recoverability, and security. In today's ever changing world of technology it is often lost on people that the level of criticality of a data center is not defined by the size of the data center. Whether a data center is 10 servers or 10,000 servers, its importance is defined by business goals and the people the business impacts. While it might seem that many technology vendors tend to focus efforts in the large data center space, the fact remains that the small/medium business (SMB) space represents a substantial part of the IT marketplace.

VMware NSX has certainly altered the landscape of network and security capabilities when it comes to data center design. With its strong features in network and security virtualization, VMware NSX offers new solutions to business problems that are being implemented with every passing day. But once again these solutions are not limited to large enterprise. The SMB markets are concerned about data center security every bit as much as the Fortune 100, and with VMware NSX there is equal access to advanced network and security for companies of all size.

This books aim is to give architects and engineers the necessary tools and techniques to transform their data center from legacy architecture to software defined architecture. The book provides a recipe of success, a well-orchestrated path to success, and a step-by-step approach to implement network and security virtualization that is proven and adopted by many in the industry irrespective of size. After all, the most important data center is YOURS!

Chris McCain
Director of Product Management, NSBU

Introduction

Before diving into a detailed discussion around designing and building NSX in SMB data centers, it is important to introduce VMware NSX and define "small data center" in the context of SMB NSX implementation.

What is VMware NSX?

VMware NSX is a complete software-based solution which has emerged as the leading product for virtualizing data center networking and security. NSX supports a broad range of core services including L2VPN (Layer 2 Virtual Private Networks), L3VPN (e.g., IPSec, SSL), NAT (Network Address Translation), and application load balancing. Businesses have deployed VMware NSX to realize the benefits of virtual networks and software defined networking (SDN). NSX has been deployed in small, medium, and large scale data centers, enabling a wide range of use cases including security, automation, and business continuity/disaster recovery.

NSX Use Cases

Figure 1.1 lists key use cases taken from production deployments of a wide cross-section of NSX customers.

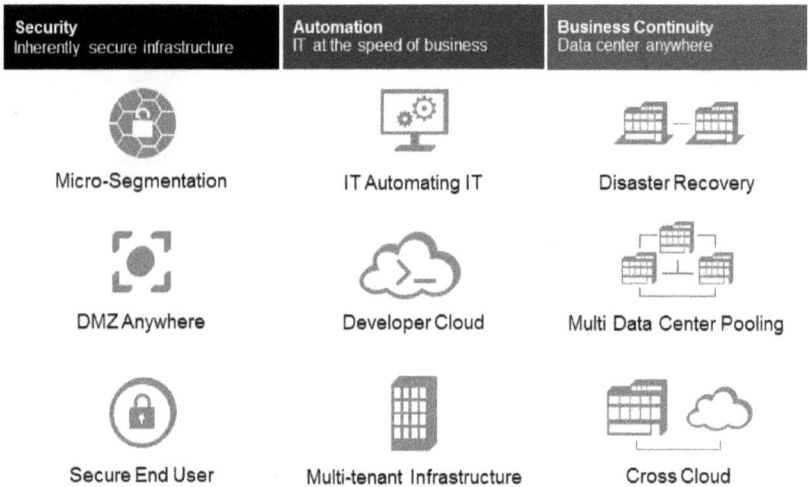

Security Inherently secure infrastructure	Automation IT at the speed of business	Business Continuity Data center anywhere
Micro-Segmentation	IT Automating IT	Disaster Recovery
DMZ Anywhere	Developer Cloud	Multi Data Center Pooling
Secure End User	Multi-tenant Infrastructure	Cross Cloud

Figure 1.1 NSX Customer Use Cases

No other network or security vendor addresses this broad range of use cases under one product umbrella. NSX, being a complete software product, is extremely flexible. It can be deployed to suit various business models and use cases that go beyond security, automation and application continuity. Regardless of the size of the data center, organizations can easily solve a variety of business challenges with NSX.

Distributed firewall, agentless anti-virus, virtual VMware NSX® Edge™ firewall, application load balancer, and VPN are attractive NSX offerings that map well to SMB demands and use-cases. The following sections examine these use cases one by one.

Security

It is no longer acceptable to utilize the traditional approach to data center security – a legacy model built around a strong perimeter defense with minimal internal protection. This model offers little protection against the most common and costly attacks targeting organizations today, including attack vectors originating within the perimeter. The ideal solution to complete data center protection is to secure every traffic flow inside the data center with a firewall, allowing only the flows required for applications to function. This ideal solution also proposes creation of micro-segments to enable a least privilege security model. These segments should be based on application needs and characteristics rather than physical constructs such as VLANs and IP addresses.

NSX micro-segmentation based security decreases the level of risk and increases the security posture of the modern data center. Micro-segmentation utilizes NSX capabilities such as distributed stateful firewalling (DFW), topology agnostic segmentation, centralized ubiquitous policy control of distributed services, network based isolation, and policy-driven unit-level service insertion and traffic steering.

Readers are encouraged to read VMware's *Micro-segmentation Day 1* Guide for further details on this topic.

Automation

VMware NSX provides a complete RESTful API to drive automation of networking, security, and services. In small data centers, automation options like REST API and PowerNSX can be useful to programmatically configure network and security services or to pull the information from VMware NSX deployments for simple operations tasks. VMware NSX automation use-case is also one of the enablers for private cloud in the data centers.

Business Continuity / Disaster Recovery (BCDR)

NSX provides ways to easily extend networking and security to up to eight instances of VMware vCenter® – either within or across data centers. This feature is commonly referred to as "NSX Across vCenters" or simply "NSX Cross-VC". NSX can extend or stretch layer L2 and L3 networks across data centers in a distributed fashion. NSX also ensures that the security policies are consistent across those stretched networks and across multiple data centers, providing a seamless, distributed, and highly available network and security overlay. All of this is done using software-based technologies rather than expensive hardware.

NSX Cross-VC is a viable option in small data centers for organizations looking to protect workloads for use cases such as disaster avoidance, disaster recovery (DR), and resource pooling.

What is the Definition of Small Data Center?

The term "small data center" is often tied to a data center implementation in an SMB. However, this strict mapping of small data center with the small and medium business (SMB) segment is not always correct.

Small data center does not always mean an implementation in a small business. A large enterprise might create multiple small data centers within its large footprint, depending business need and operational feasibility. These small data centers could also be viewed as demarcation points between distinct teams or business functions.

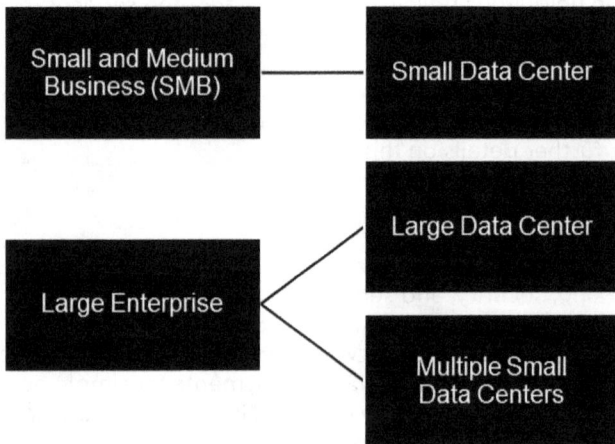

Figure 1.2 Small Data Center(s) within SMBs and Large Enterprises

This discussion should make it clear to the reader that small data center is a very loosely defined term. For a more inclusive scope, consider the following concepts for framing a typical small data center:

- Fewer than 20 ESXi hosts

- All ESXi hosts in a single cluster

- North-south bandwidth requirements for workloads do not exceed 10G

Chapter 1 - Key Takeaways

- NSX footprint scales with the business requirement and needs.

- NSX scale is not tied to specific hardware or size.

- NSX is widely deployed by small, medium, and large enterprises and service providers.

- Small and medium business have deployed NSX to enable many use cases in the areas of security and network services such as Edge firewall, load balancing, NAT, and VPN.

- The small data center concept is not always tied to SMB.

- Small data center could be one or more independent pods serving a specific function or application within a large enterprise.

NSX for vSphere

Rise of SDN / Network Virtualization

Before examining NSX's components and software-defined approach toward networking and security, first reflect on traditional networking and security methodology.

In a traditional networking model, the management, control, and data planes are combined together into a single physical device (e.g., router, switch, or firewall). Distinct CLI (Command Line Interface) or browser based UI (User Interface) access is required to manage each device. The configuration is pushed only to that single device where the control and data planes are programmed. This has to be done on each and every physical device in the network. While the industry has evolved over time, more recently providing element managers to ease the management and configuration of devices, these are still essentially hardware-based solutions. They do not provide the adaptive load distribution, reduced failure domain, flexibility, automation, or agility that is demanded by modern data centers.

NSX follows a true software-defined networking approach where the management, control, and data planes are separated from each other. This model allows networking and security services to be faithfully reproduced in software – providing agility, automation, and resiliency. This is what organizations are demanding to modernize their data centers.

Importance of a Solid vSphere Design

vSphere is the foundation for NSX deployment. When building a house, the foundation plays an important role and serves as the base for the structure; it is the first thing to be implemented, and if the foundation is not solid, the rest of the house will always have an underlying weakness. Similarly, for a successful NSX deployment, it is imperative to have a solid vSphere design and implementation in place. This is needed regardless of the size of the data center.

vSphere design discussion is outside the scope of this book. Readers should be aware that at minimum, a solid vSphere design requires appropriate vSphere clustering, careful consideration for the VMware vSphere® Distributed Switch™ (VDS), and proper implementation of vSphere functionality. It also requires properly sized compute, network, and storage resources. For detailed discussions on these topics, refer to the following design guides:

- VMware NSX Reference Design Guide
  https://communities.vmware.com/docs/DOC-27683

- VMware Validated Design Guides
  https://www.vmware.com/solutions/software-defined-datacenter/validated-designs.html

NSX for vSphere Components

Figure 2.1 shows the various layers of NSX architecture based on the role performed by each component. At a high level, the NSX solution architecture is divided between management, control, and data planes. This approach provides the advantage of decoupling NSX from hardware dependencies, allowing all networking services to be virtualized following the same operational model that compute and storage virtualization has provided for years.

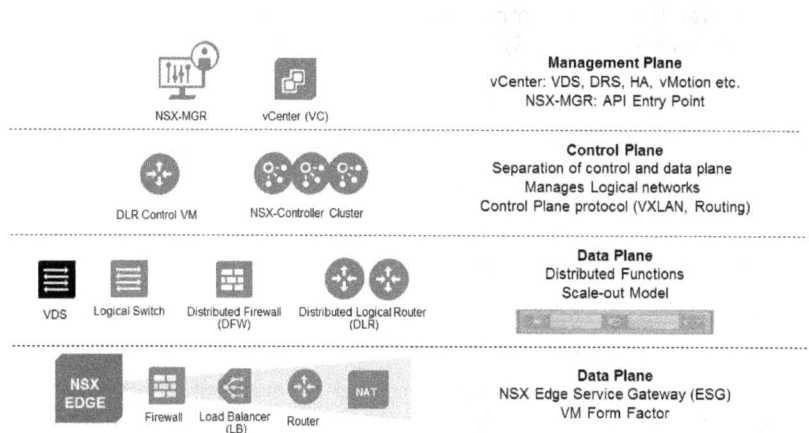

		Management Plane vCenter: VDS, DRS, HA, vMotion etc. NSX-MGR: API Entry Point
NSX-MGR	vCenter (VC)	

Management Plane
vCenter: VDS, DRS, HA, vMotion etc.
NSX-MGR: API Entry Point

DLR Control VM   NSX-Controller Cluster

Control Plane
Separation of control and data plane
Manages Logical networks
Control Plane protocol (VXLAN, Routing)

VDS   Logical Switch   Distributed Firewall (DFW)   Distributed Logical Router (DLR)

Data Plane
Distributed Functions
Scale-out Model

NSX EDGE   Firewall   Load Balancer (LB)   Router   NAT

Data Plane
NSX Edge Service Gateway (ESG)
VM Form Factor

Figure 2.1 NSX for vSphere Components

Management Plane

VMware vCenter and NSX Manager are the minimum required components for the NSX management plane. There is a one-to-one relationship between each vCenter and NSX Manager instance. NSX Manager handles configuration management for virtual network and security services along with serving as REST API entry point.

The NSX Manager VM can be deployed along with other management/operations VMs and can coexist with VMware (e.g., vRealize Log Insight, vRealize Operations, vRealize Network Insight) or third party (e.g., Splunk, Gigamon) offerings. For the sake of simplicity, other management and operations VM are not shown in Figure 2.1.

Control Plane

The control plane manages logical networks and is responsible for programming VXLAN and routing information within the ESXi hypervisors. The NSX controller maintains the runtime state of the virtual network by maintaining and updating MAC, ARP, and VTEP tables. Regardless of the size of the NSX deployment, there must be three NSX controller VMs created. These three NSX controller VMs form their own cluster. (Note - it is important to understand the purpose and difference between the vSphere cluster vs the NSX controller cluster.)

Data Plane

The data plane is divided into two layers:

1. Data plane running as distributed function

2. Data plane running in the ESG VM(s)

Distributed functions such as logical switching and routing are programmed directly in the ESXi hypervisor kernel. The distributed nature of these functions provides a scale-out architecture that offers greater performance, resiliency, and high availability when compared with traditional networking models.

NSX Edge Service Gateway (ESG) is a virtual appliance that is also in the data path. As shown in Figure 2.2, the ESG VM sits at the edge of the virtual network and can be considered an on-ramp/off-ramp demarcation point between physical and virtual networks. The ESG VM can provide edge services such as firewall, L2VPN, L3VPN, load balancer, and edge routing.

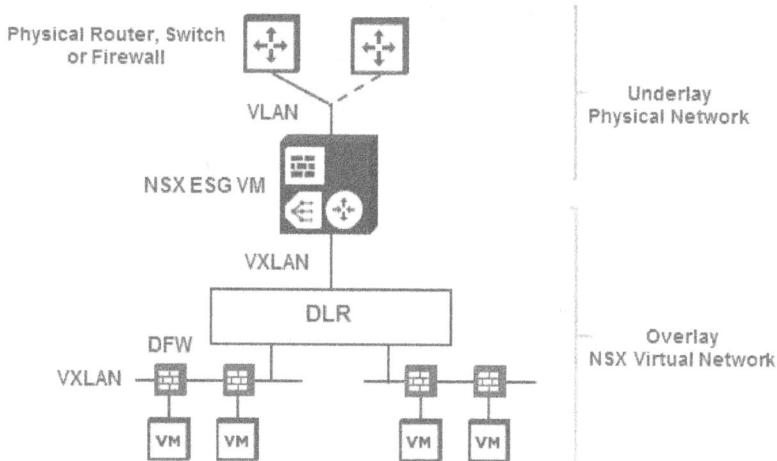

Figure 2.2 NSX ESG VM providing connectivity between physical underlay and virtual overlay networks

NSX in Small Data Center Use-Cases

SMB data center use cases can be divided based on business function and application-specific characteristics.

Functional Level Use Cases

Organizations deploy NSX with a small footprint in specific functional areas or groups. Examples include:

- Disaster recovery / disaster avoidance

- Pre-production / test environments

- Compliance / DMZ

- Business unit specific operational models

- Micro-segmentation

Application Level Use Cases

Many enterprises deploy NSX in small data centers to tackle one or more application level use cases:

- VDI

- Load balancer

- Agentless antivirus

NSX Advantage for Small Data Centers

Organizations adopt NSX because of its technical strengths and the advantages of deploying networking services in software. They also benefit in terms of simplicity, support, and ease of operations.

Simplicity and Modularity

Small enterprises like the idea of its simplicity and modularity. Organizations can have peace of mind to grow and add more features as they increase their capacity or user base. There is no need to purchase all the networking hardware up front with lots of unknowns down the road. These unknowns could be:

- Hardware TCAM exhaustion due to increase number of VMs

- Change in the technology that could result in replacing hardware

- Network design changes requiring fewer routers, switches, and firewalls than originally planned

NSX provides businesses a software-based networking service that can be spun up or torn down at any time without incurring additional hardware costs.

Single Support Call

Organizations appreciate that networking and security services are bundled within the same product. They do not need to worry about contacting multiple vendors for service or support.

Ease of Operations

A majority of enterprises are already familiar with the vSphere operational model. With vSphere, VMs and storage are abstracted in software, removing the physical dependencies of servers or storage arrays. These VMs can be created or destroyed in seconds. VMs can move on-demand or through automation anywhere in the data center based on resource availability and utilization.

NSX seamlessly integrates within this same operational model, minimizing the learning curve for adoption and operation. Furthermore, it enhances the environment by allowing operators to create networks and networking services – both on-demand or through automation – using similar processes to those for servers and storage.

The next chapter discusses the importance of well architected vSphere design. vSphere design – including but not limited to vCenter, Platform Services Controller™ (PSC), vSphere Cluster, VDS, DRS, and HA – is the foundation for a successful NSX deployment. These aspects must be reviewed carefully prior to designing NSX based overlays, virtual networks, and services.

Chapter 2 - Key Takeaways

- NSX can run on any hardware underlay technology that provides robust, redundant, and highly available IP connectivity.

- A resilient and solid vSphere design is the foundation for a successful NSX for vSphere deployment.

- NSX decouples management, control, and data planes, running them in a virtual and/or distributed manner – all in software.

vSphere Cluster Design and NSX Deployment Models

The vSphere cluster design plays an important role for any type and size of NSX deployment. Before going into the details of small data center deployment models, it is important to understand and appreciate the need for a solid vSphere cluster design. It is also vital to understand why multiple vSphere clusters are created in large data center designs and why those design choices are not applicable to a small data center.

Large Data Center Cluster Design

For large data center design, vSphere deployment best practices recommend creating purpose-built vSphere clusters. NSX design guides and deployment best practices also suggest creating separate vSphere clusters for management, edge, and payload virtual machines as well as distributed functions. As shown in Figure 3.1, purpose built clusters not only limit the fault domain but also help carve out resources based on the functions that these clusters perform.

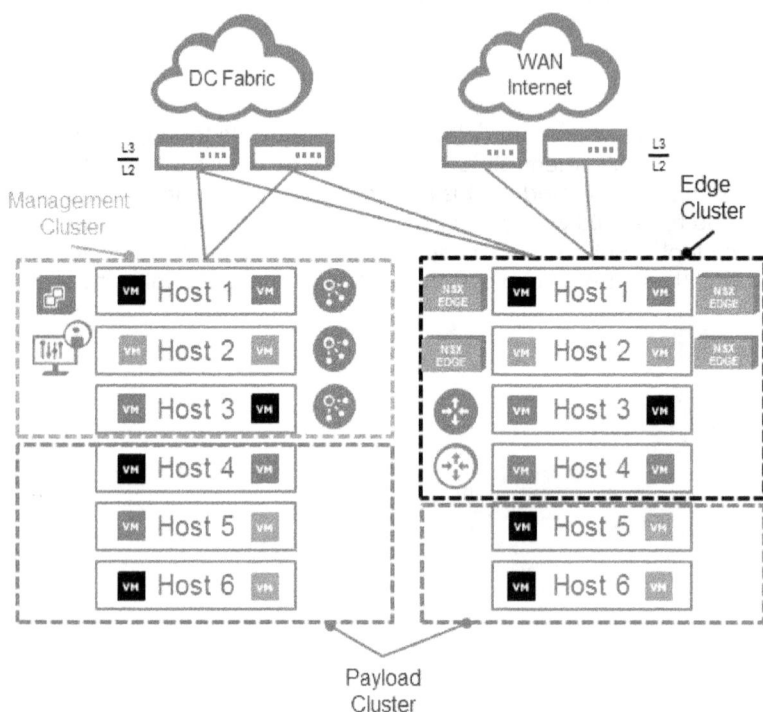

Figure 3.1 Large Data Center Cluster Design

Small Data Center Cluster Design

In large data centers with many ESXi hosts, the recommendation for purpose-built vSphere clusters can easily be justified and achieved. In small data centers where there are few ESXi hosts available, businesses can deploy management, compute, and edge workloads into a single vSphere cluster. Figure 3.2 shows a typical small data center implementation where all NSX and vSphere components as well as payload VMs are deployed in a single cluster to provide rich set of NSX networking and security services.

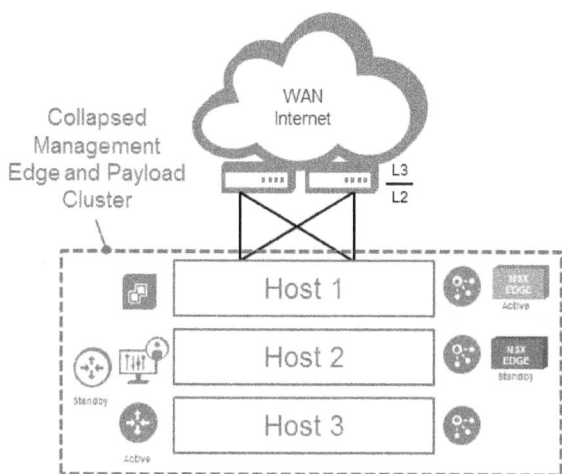

Figure 3.2 Typical Small Data Center Collapsed Cluster Design

Deploying NSX in a single cluster requires proper resource reservation, sizing, and monitoring of management, edge, and payload VMs. This includes vCenter, NSX Manager, NSX controllers, DLR control VM, ESGs, and service VMs. These components should be carefully allocated and monitored so that as the environment changes, adequate resources are maintained. Lack of sufficient resources could cause a degraded experience for both critical and non-critical business applications.

Small Data Center Deployment Models

NSX is an extremely flexible and extensible solution. It can be implemented in various deployment models, with specific approaches popular among small and medium businesses. These models map closely to core applications, business demands, and growth patterns. These options are discussed below and will be detailed throughout this book:

- Security focused deployment model

- Centralized Edge deployment model

- Full stack deployment model

Each of the models is summarized below and will be discussed in detail later in this chapter.

Security Focused Deployment Model

The security focused deployment model is the most popular and widely deployed model, not only in small and medium businesses, but also in large enterprises. Businesses looking to secure applications and assets deploy this model to achieve true micro-segmentation. The key technology behind this deployment model is NSX Distributed Firewall (DFW) functionality coupled with NSX's centralized management of security policies. NSX DFW is a scale-out, distributed, and stateful firewall technology embedded at the hypervisor kernel.

This model has become the starting point in majority of the NSX deployments.

Centralized Edge Deployment Model

The centralized Edge deployment model is based on the ESG VMs. All the traffic entering or leaving the virtual network (i.e., north-south) or traffic within the data center (i.e., east-west) passes through this ESG VM. Businesses that opt for this model usually do so because they are not ready to implement VXLAN in their networks. This could be due to business/operations priorities or challenges with changing the network MTU (Maximum Transmit Unit) for VXLAN (Virtual Extensible LAN) enablement.

This model could be used as a transition in situations where an organization is moving from a security focused deployment model to a full stack deployment model but is not read to fully implement a network overlay.

Full Stack Deployment Model

The full stack deployment model helps businesses realize the full benefits of software defined virtual networking by implementing features such as distributed logical switching, distributed logical routing, and layer 2 bridging. In this model, enterprises create network overlays and offer various networking services in a software-only model. The general guidelines from VMware are to deploy NSX using the full stack deployment model.

The emphasis of this book will be on security focused and full stack deployment models ; these are the two models most widely adopted. It is important to understand that the order in which these models are presented does not have bearing on how they should be implemented.

Security Focused Deployment Model Details

Security focused deployment is the most popular choice for small and medium businesses. This model is also the foundation of micro-segmentation and allows applications to be segmented based on security requirements rather than traditional network constructs like IP addresses. The security focused deployment model can be further divided into following sub-models:

- Security focused deployment model with distributed firewall

- Security focused deployment model with agentless anti-virus and anti-malware

- Security focused deployment model with deep packet inspection

As evidenced in these examples, the security focus model goes beyond the DFW use case. NSX provides a holistic policy-based approach and framework to enable end-to-end data center security. Businesses can easily insert security services like agentless antivirus into this model without redesigning or disturbing an existing DFW implementation. An agentless AV solution is delivered via the NSX guest introspection framework.

Besides guest introspection, the network introspection (e.g., IPS/IDS, L7 inspection) service can be inserted into the NSX security focus deployment model. Network introspection is not a popular choice in small data centers; businesses are more inclined to deploy DFW with guest introspection services and then deploy network introspection services with an NSX full stack model.

All of the above models can be deployed on their own or alongside other models. These deployment models offer different design patterns based on the specific use cases driving the micro-segmentation requirements in data centers.

Security Focused Deployment Model with Distributed Firewall Use-Case

Figure 3.3 shows a logical topology of a security-focused deployment using DFW. Deployment of DFW is non-disruptive and does not require any changes to the physical infrastructure such as MTU or routing. It also does not require VXLAN functionality. Large enterprises that deploy small data centers or self-contained, function-based racks (e.g., DMZ, VDI, QA/dev) can also take advantage of this model. Certain NSX features – such as security tags and dynamic, application centric security groups – are popular in the SMB space.

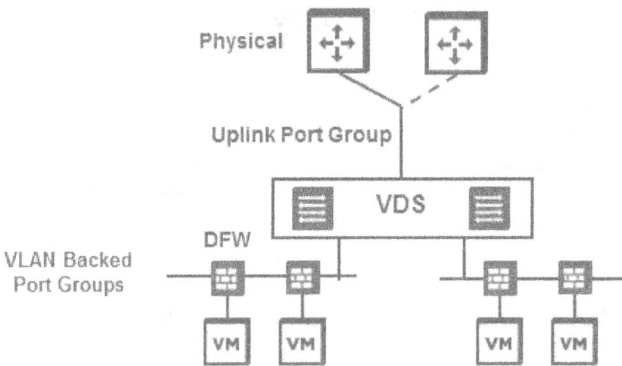

Figure 3.3 Security Focused Deployment Logical View

vSphere Distributed Switch Requirement

vSphere Distributed Switch is the only virtual switch supported for NSX; hence, it is required for micro-segmentation /DFW. Organizations who have deployed payloads behind a vSphere Standard Switch (VSS) must migrate them before deploying NSX.

VM Footprint

The footprint necessary to implement an NSX security focused deployment model with DFW is extremely minimal. In addition to the vCenter and the Platform Services Controller, the NSX Manager VM is the only additional VM needed to implement distributed firewall. Businesses can easily deploy it on two ESXi hosts with vCenter/PSC and NSX Manager VMs. It is recommended to have at least three hosts in production so there is extra ESXi available in case of host failure. Table 3.1 shows an example VM footprint where an organization deployed this use case using only two virtual machines. Note that vCPU, memory, and storage values are from vCenter 6.0 and NSX 6.3.3, the versions used throughout this book.

Table 3.1 Minimum VM footprint for Security Focused Deployment Model with DFW

VM Name	vCPU	MEM (GB)	Storage (GB)	VM Count
Tiny vCenter Appliance with Embedded PSC	2	8	116	1
NSX Manager	4	16	60	1
Total	6	24	176	2

vSphere Cluster Layout and VM Placement

Figure 3.4 shows the suggested cluster layout and VM placement. NSX Manager, vCenter, PSC, other management related VMs and payload VMs are collapsed in a single vSphere cluster due to the smaller number of ESXi hosts available. As there is no ESG VM deployed, there is no concept of an Edge cluster in this model. Since this model does not require VXLAN, the payload VMs will be behind VLAN backed-port groups to implement DFW.

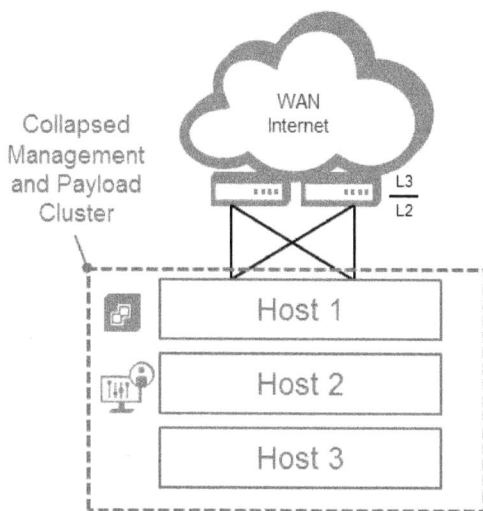

Figure 3.4 Security Focused Deployment with DFW

Note that Figure 3.4 does not show the requirement for payload VMs. In all designs, the payload VMs are separate from infrastructure VMs and they vary based on specific application needs. Businesses should size according to distinct workload requirement and add additional compute hosts as necessary. It is always advisable to have more hosts than are minimally required to address a host failure.

No Need to Deploy the NSX Controllers

This model does not require NSX Controllers or Edges, further helping organizations reduce the VM footprint in small data centers. In some situations, an organization may want to deploy NSX controller in advance if it has future plans to use VXLAN and distributed routing.

Security Focused Deployment Model with Agentless Anti-Virus and Anti-Malware Use Case

NSX offers an agentless AV feature through integration with certified 3rd party vendors including Trend Micro, McAfee, and Symantec. This integration is offered via the NSX guest introspection framework.

The guest introspection requires additional guest introspection service VMs and partner service VMs per ESXi host in the vSphere cluster. These services VMs (SVMs) are automatically deployed at the vSphere cluster level. SVMs are automatically deployed when a ESXi hosts enters the cluster and automatically removed when an ESXi host leaves the cluster.

Figure 3.5 shows the design layout for this use case in a single cluster. Notice the addition of the NSX guest introspection SVM and partner SVM.

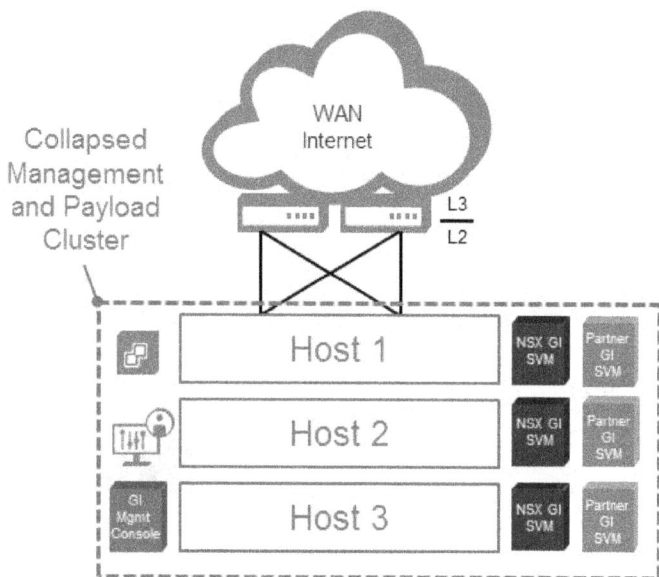

Figure 3.5 Security Focused Deployment with Agentless-AV

Service VM Mobility

The SVMs are tied to one particular host when deployed. These SVMs should not be vMotioned or Storage vMotioned to any other hosts.

Security Focused Deployment Model with Deep Packet Inspection (IPS/IDS) Use Case

NSX offers layer 7 deep packet inspection capabilities through integration with certified 3rd party vendors such as Palo Alto Networks, Trend Micro, McAfee, Fortinet, and Symantec. This use case requires a partner network introspection service VM for each ESXi host. These services VMs are automatically deployed when an ESXi hosts enters the cluster and automatically removed when an ESXi host leaves the cluster.

This model can be deployed with or without the agentless-AV use case. Figure 3.6 shows a mode where all security-focused services are deployed in a single cluster.

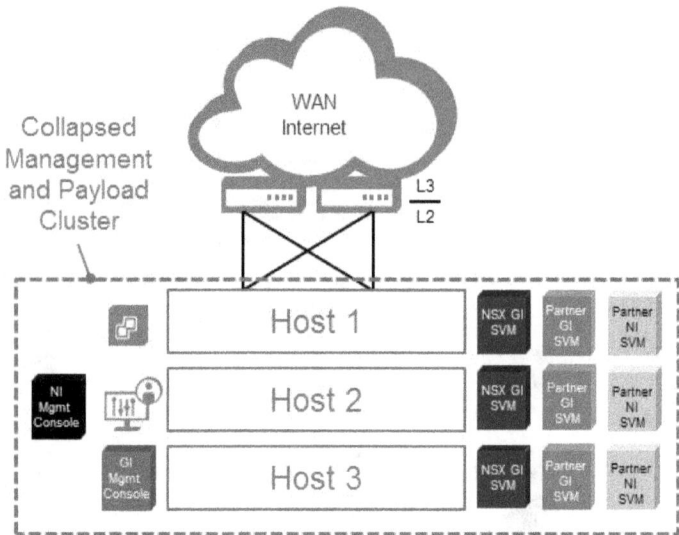

Figure 3.6 Security Focused Deployment with Agentless AV and Deep Packet Inspection

Centralized Edge Deployment Model

The NSX Edge Service Gateway VM is the core component in this deployment model. An ESG VM can be deployed with or without NSX DFW. This deployment model could be adopted as an intermediate step in going from a security-only deployment to an NSX full stack deployment with VXLAN or DLR.

For a small business, this model is attractive due to its low VM footprint and simple deployment with minimal changes to the physical infrastructure. It can serve as an additional layer of defense for a DMZ in a small data center by providing virtual perimeter edge firewall capabilities.

Design Considerations

From the design point of view, it is important to notice that this model does not provide traffic optimization; all the traffic has to go through the NSX ESG VM.

MTU Size

The centralized Edge deployment model does not mandate implementing VXLAN. Employing VXLAN with the centralized Edge model is optional; this variation will be discussed in an upcoming section. Since there is no requirement to implement VXLAN, no MTU changes are required on the infrastructure.

This deployment model can be implemented with existing VLAN backed port groups. This is important for SMBs as it allows continued use of their existing operational models.

Consider increasing the MTU to jumbo size in green field deployments to provide the greatest flexibility for the full stack deployment model support going forward.

Default Gateway

It is recommended to use the ESG VM as the default gateway for the payload VMs. Note that all the traffic will be hair-pinned, even where the payloads are sitting next to each other on the same ESXi host. Having the ESG VM as the default gateway will also help when the business is ready to move to a full stack deployment model. At that point, the default gateway IP address could be reassigned to the DLR.

Edge Services

The NSX ESG is a multi-function gateway in VM form factor. It supports multiple services on a single ESG, including dynamic/static routing, firewall, NAT, load balancer, and L2/L3 VPN. The sizing of an ESG varies based on the services configured. For the majority of use cases, the large VM form factor is sufficient. In the case of an ESG with an L7 load balancer, the extra-large form factor is recommended. An ESG VM is involved in both control and data traffic.

ESG VM Connectivity

Figure 3.7 shows a logical topology where an NSX ESG VM is connected to physical device on the north side (i.e., external link). The ESG VM is typically connected to an uplink switch using single NIC. This offers a simple and easy to manage deployment. The southbound (i.e., internal) link or links are connected to the dvPortGroup where the payload/workload VMs are connected. In essence, both northbound and southbound links are VLAN backed port-groups. In small data centers, it is highly recommended to deploy ESG in active/standby HA mode to provide resiliency in case of an active ESG VM failure. NSX Edge HA and vSphere technologies (e.g., vSphere HA, DRS) also help improve the overall availability in this deployment model.

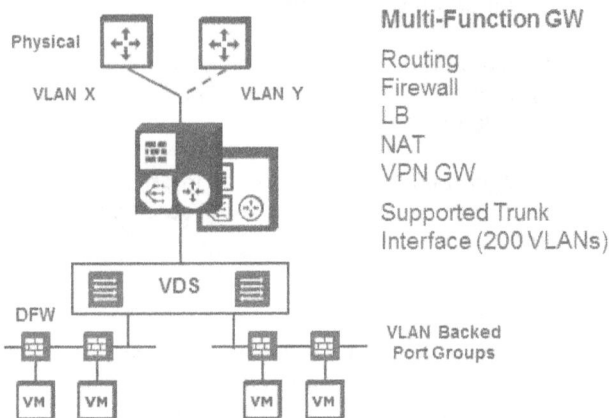

Figure 3.7 Centralized Edge Deployment Model

vSphere Cluster Layout and VM Placement

Figure 3.8 shows a possible placement of VMs on a single vSphere cluster with three ESXi hosts.

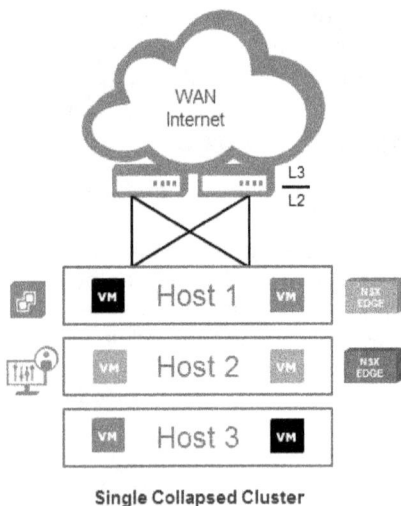

Single Collapsed Cluster

Figure 3.8 VM Placement for Centralized Edge Deployment Model

In this layout, the NSX Edge VM is deployed in HA. It is important to make sure that both ESG VMs are not placed on the same host; in the case of host failure, both VMs would go down, impacting application accessibility. When DRS is enabled on the vSphere cluster, the proper anti-affinity rule is created automatically at the time of ESG VM creation with HA configuration. This is done via NSX's built-in automation to provide protection against single host failure.

Centralized Edge Deployment Model with VXLAN Logical Switches

This model is applicable in situations where workloads are switched only within distinct L2 domains and cross-rack east-west routing optimization is not required. In this case, deploy applications behind VXLAN backed port groups – also known as VXLAN logical switch or simply logical switch – to take advantage of seamless L2 extension across racks without requiring NSX DLR for east-west traffic optimization.

Deploying applications and workloads behind a logical switch also eases migration to a full stack deployment. This model does require deployment of NSX controllers, though the MTU must be at least 1600 bytes for the VXLAN transport VLAN. The default gateway for workload/application VMs will be the internal interface of the ESG. When the business is ready to deploy DLR and NSX in full stack mode, they can easily switch the default gateway from the ESG to the DLR with minimal intervention.

In this model, all cluster functions – management, edge, and payload – could reside on a single vSphere cluster. The entire cluster will be prepared for NSX, ensuring that DFW and VXLAN will be available on all hosts. Refer to the "Full Stack Deployment Model " section for additional design consideration with central Edge with VXLAN logical switch model.

Full Stack Deployment Model

The full stack model provides complete abstraction from underlying hardware by utilizing industry standard VXLAN technology. It allows businesses to deploy networking topologies in software, eliminating the dependencies and constraints of physical hardware. This model not only provides features such as distributed firewall but also optimizes east-west routing with the NSX logical switch and distributed logical router. With this model, businesses get the benefit of making security and routing decisions close to the workload. Figure 3.9 shows a full stack deployment model logical topology showcasing these various features.

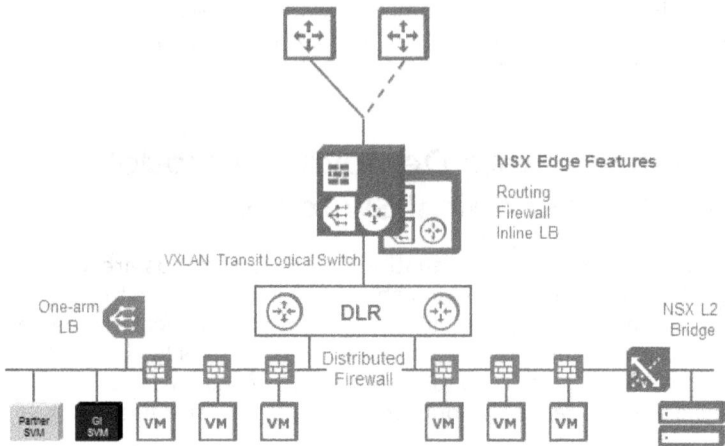

Figure 3.9 Full stack deployment model logical topology

This design is also popular for disaster avoidance and disaster recovery scenarios where a business wants to recover a portion of their main site to a small data center location.

Design Considerations

This model requires an MTU of at least 1600 bytes on the physical infrastructure. This requirement to increase MTU size only applies to VLANs used for VXLAN Tunnel Endpoint (VTEP) vmkernel interfaces on ESXi hosts. Additionally, robust IP connectivity is required between VTEP interfaces.

Table 3.2 Various functions and components in full stack deployment model

Function	Components
Management Plane	NSX Manager, vCenter and other management VMs
Control Plane	NSX controller and active/standby DLR control VMs
Data Plane East-West	ESXi host kernel modules (VDS, DFW, VXLAN, DLR)
Data Plane North-South	ESG VMs (HA or ECMP Mode)

Table 3.2 shows VMs and ESXi hypervisor kernel modules needed to design and deploy a full stack deployment model.

Deployment Considerations

In this model all VMs (e.g., management, edge, and payload) reside on a single vSphere cluster. The entire cluster will be prepared for NSX, providing DFW, VXLAN, and DLR for all hosts.

It is mandatory to have at least three ESXi hosts for the deployment as full stack model requires an NSX controller cluster. These NSX controller VMs should be running on separate hosts; create anti-affinity rules to ensure this policy is enforced.

It is recommended to at least have 4 ESXi hosts for a production deployment, providing resiliency in case of single host failure. Figure 3.10 shows a possible placement of VMs to deploy the full stack model. In this particular example, static routing is used between NSX DLR and NSX Edge ; hence there is no DLR control VM shown. Also notice that only one NSX Edge VM is active in this example because the NSX ESG is deployed in HA mode.

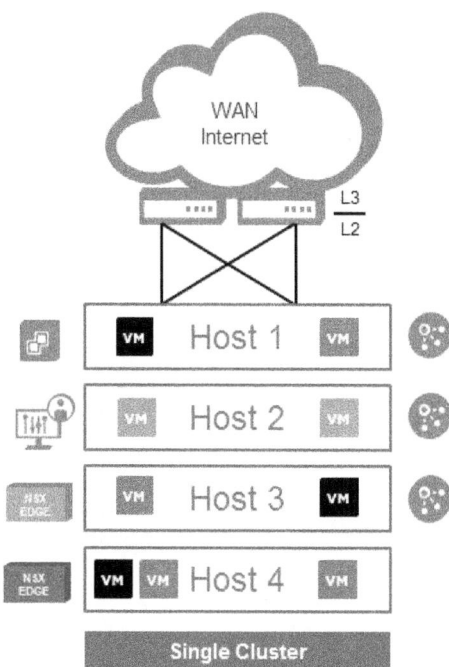

Figure 3.10 Typical VM placement in single vSphere cluster for full stack deployment model

Table 3.3 can be used as a reference to calculate the resources needed to deploy VMs in this model. Notice that this does not include requirements for workload VMs. Payload/workload VM requirements should be calculated and taken into consideration in addition to the NSX component VMs and is outside the scope of this book.

Dynamic routing (e.g., BGP, OSPF) is also supported in this model. Dynamic routing between a DLR and ESG will require deployment of a DLR control VM in HA, increasing the number of VMs by two. DLR control VMs are by default deployed as a compact VM and do not need to be resized. ESG VMs come in different form factors, though it is recommended to use large form factor for a majority of production deployments.

Table 3.3 shows the VM footprint needed to deploy a full stack deployment model in a small data center. In this particular deployment, the organization is using a tiny vCenter appliance and an ESG in HA mode with static routing.

Table 3.3 Minimum VM footprint needed to deploy full stack deployment model
* ESG in HA model with static routing

Component	VMs	vCPU	MEM (GB)	Storage (GB)
Tiny vCenter Appliance with Embedded PSC	1	2	8	116
NSX Manager	1	4	16	60
Controllers	3	12 (3x4)	12 (3 x 4)	60 (3x20)
Edge VM (Large)*	2	4 (2x2)	1 (2 x 0.5)	2 (2 x -1)
Total	7	22	37	~ 238

Chapter 3 - Key Takeaways

- NSX can be deployed in production in small data centers with a single vSphere cluster.

- The most popular deployment models are:

  - Security Focused Deployment

  - Full Stack Deployment

- Organizations that started with security focused (i.e., micro-segmentation) deployment quickly moved to full stack deployment because of the benefits and advantages NSX brings to the table.

Individual Component Design and Deployment Considerations

The focus of this chapter is on the design and implementation aspects of the key NSX and vSphere components. The features provided by both NSX and vSphere (e.g., clustering, DRS, affinity rules) make the NSX deployments robust and resilient for the production workload. With the appropriate use of these options, businesses have been able to successfully deploy NSX with high availability. Accurately sizing these component is important to meet business requirements and SLAs.

vCenter Server

For NSX for vSphere deployment, vCenter and vSphere Distributed Switch are prerequisite components. This book does not go into the details of designing vSphere and vCenter; it is assumed that vCenter is already deployed with vSphere deployment best practices.

vCenter can be deployed in tiny, small, medium, and large configuration. Table 4.1 illustrates the number of hosts and VMs that are supported per vCenter deployment type. Table 4.1 also shows a potential mapping between VC deployment options and NSX deployment types.

Table 4.1 VC Deployment Options

VC VM Options	Max Number of Hosts	Max Number of VMs	Potential NSX Deployment Type	vCPU	MEM (GB)	Disk (GB) Embedded PSC
Tiny	10	100	Lab/PoC	2	8	116
Small	100	1000	Small DC	4	16	136
Medium	400	4000	Medium DC	8	24	275
Large	1000	10,000	Large DC	16	32	325

For small data center design, a business will typically deploy a small vCenter VM. Consider deploying medium or large vCenter VM if there are plans to grow the number of hosts or maximum number of VMs.

vSphere License

There are different types and levels of vSphere licenses available for different businesses and use cases. Table 4.2 lists the three most popular vSphere license types and the supported sizing and features associated with them. Coverage of each and every licensing option is outside the scope of this book.

VDS is one of the main requirements for NSX deployment, and is available with the vSphere Enterprise+ license. With the purchase of the NSX license, VDS is available without the Enterprise+ license requirement; however, DRS is extremely important for NSX deployment and is available only with the Enterprise+ license. Procurement of the Enterprise+ license is highly recommended.

DRS with its anti-affinity rule ensures that each of the three NSX controller VMs reside on a different host. In a situation where appropriate DRS rules were not created for an ESXi host running two NSX controller VMs, a failure of that host will bring both NSX controllers down. This will put the NSX controller cluster in read-only mode.

Table 4.2 vSphere License

vSphere License Options	Notable Features/Limitations
Essential+	Supports up to 3 hosts, standard vSwitch and vSphere HA
Standard	Supports up to 1000, standard vSwitch and vSphere HA
Enterprise+	Supports all enterprise grade DC virtualization features (HA, DRS, VDS etc.)

NSX License

There are three different licensing tiers for NSX. Table 4.3 contains a partial list of NSX features available by license types.

Table 4.3 NSX license tiers and corresponding features

Features	Standard	Advance	Enterprise
Distributed Routing and Switching (DLR/ VXLAN)	✓	✓	✓
NSX ESG (except load balancer)	✓	✓	✓
SW L2 bridging	✓	✓	✓
Distributed Firewall (DFW – Micro-Segmentation)		✓	✓
NSX Edge load balancing		✓	✓
Cross vCenter NSX			✓

The features and options mentioned in Table 4.3 are a subset of what is available. More details and various licensing option are available on VMware web site at following URL:

http://www.vmware.com/products/vsphere.html

vCenter Deployment Options

The vCenter VM requires presence of the Platform Services Controller service. The PSC could be embedded as part of the vCenter VM itself or deployed as a separate VM. The most popular deployment option for small data center is to deploy embedded PSC with vCenter appliance as shown in Figure 4.1.

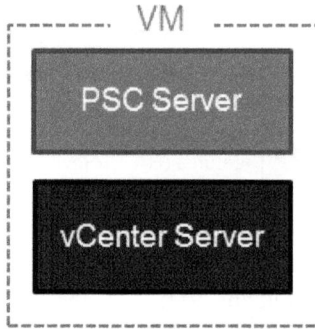

Figure 4.1 PSC and vCenter Servers embedded into a single VM

External PSC deployment option, shown in Figure 4.2, is an alternate option which is recommended for medium to large environments. It decouples the vCenter server from the PSC server, making it a more scalable option. This model allows organizations to grow the number of virtual machines without re-architecting their vSphere deployment.

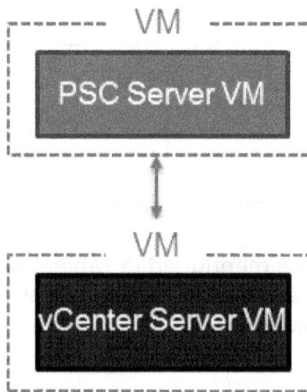

Figure 4.2 PSC and vCenter Servers running in separate VMs

NSX Manager

The NSX Manager virtual machine deployment is provided via an OVA file available from the VMware website. This single OVA bundles the NSX Manager VM along with all components and VMs needed for an entire NSX deployment. This list includes:

- NSX Manager

- ESXi VIB (vSphere Installation Bundle) files for distributed functions such (e.g., distributed firewall, logical switching, distributed routing)

- DLR control VM

- Edge Services Gateway VM

The deployment of these components and VMs is an automated process driven from the NSX Manage UI.

The NSX Manager VM is not a locked-down VM; the vCenter UI allows modification of its vCPU and memory values, as show in Figure 4.3.

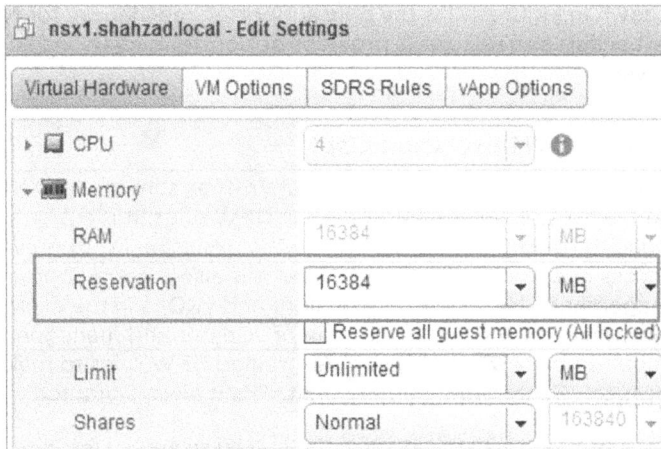

Figure 4.3 NSX Manager Memory Reservation

It is not recommended to alter these values, as changing the NSX Manager VM settings may lead to errors, disruptions, or poor performance.

NSX Manager is the management plane component; it never participates in the data path. If the NSX VM goes out of service, there is no data plane impact for existing flows – everything will keep working but no new changes can be made. It is a best practice to schedule NSX Manager backup based on an organization's recovery point objectives.

In the event of an NSX Manager VM failure or corruption, the NSX Manager should be restored from the backup on a new NSX Manager VM with the same version as the failed one.

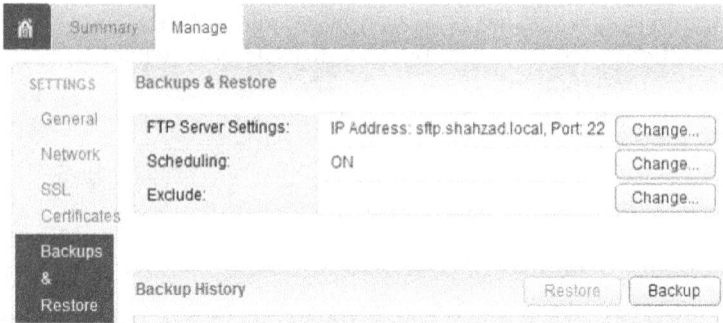

Figure 4.4 NSX Manager Backup and Restore UI

The UI shown in Figure 4.4 is the only supported mechanism for NSX Manager backup and restore. It provides options to backup via FTP or SFTP server.

NSX Manager "VM Exclusion List"

The NSX Manager VM exclusion list feature makes sure that any VM placed in the list will never be affected by DFW rules. It is recommended to add the vCenter VM to the exclusion list to prevent losing access due to a misconfigured rule. It is also a good approach to add management VMs such as Log Insight and vROps in the exclusion list as shown in Figure 4.5. If protection of vCenter and management VMs is required by NSX, configure fine-grained DFW rules to make sure access to vCenter and management VMs is always granted.

Note that NSX VMs, like NSX-MGR, DLR control VM, and NSX ESG VMs are automatically part of exclusion list.

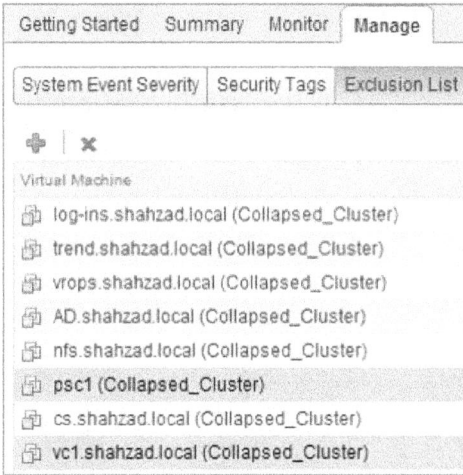

Figure 4.5 NSX Manager Exclusion List

NSX Controllers

The NSX controllers are responsible for programming the data plane (i.e., ESXi hosts) with the logical switch and distributed routing information in the kernel. NSX supports configuring three controller VMs in order to provide high availability and load distribution. NSX controller clustering, when used in conjunction with vSphere high availability and DRS, provides protection in case of failure and ensures operations are running smoothly.

In order to prevent control plane failures, creation of DRS anti-affinity rules is recommended. These rules should ensure that each controller VM resides on a separate host. It is suggested to create "SHOULD" anti-affinity rules. "SHOULD" rules are best effort rules that will put VMs on separate hosts. For a three host cluster, in case a host is not available, these "SHOULD" rules would then place two controller VMs on a single host. Use of at least 4 ESXi hosts will ensure that the cluster can sustain a single host failure.

VM/Host Rules

| Add... | Edit... | Delete |

Name	Type	Enabled	Conflicts	Defined By
📋 Controllers-Anti-Affinity-...	Run VMs on Hosts	Yes	0	User

VM/Host Rule Details

Virtual Machines that are members of the VM Group should run on hosts that are members of the Host Group.

| Add... | Remove | | Add... | Remove |

NSX-Controllers-VM Group Members		Collapsed_Cluster_Hosts Group Members
🖥 NSX_Controller_d74ab4cc-a256-412a-b5a8-9450b2f23e8f		📇 10.29.12.173
🖥 NSX_Controller_65983eea-1f36-49a5-bb7c-de6e40211f68		📇 10.29.12.170
🖥 NSX_Controller_3f42ba6b-936a-47e1-9a96-3d0a2b51ce...		📇 10.29.12.171
		📇 10.29.12.172

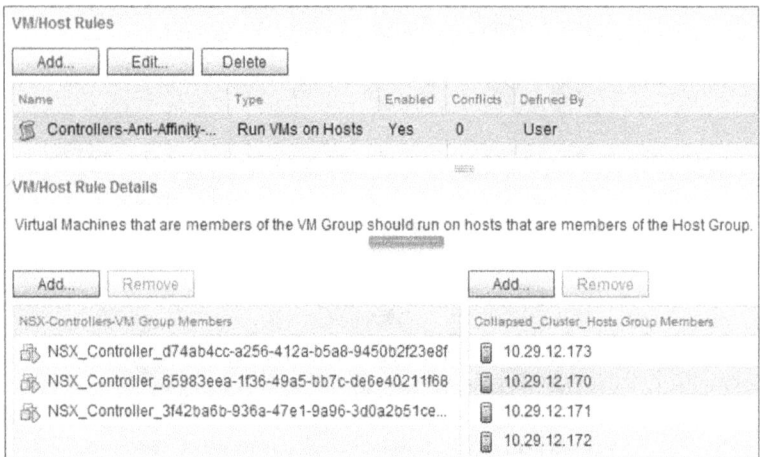

Figure 4.6 NSX Controller Affinity Rules

NSX controller VMs are deployed via the NSX Manager UI. These VMs are deployed locked-down, meaning it is not possible to change the default vCPU and memory settings. These VMs are also created with default memory reservation which cannot be changed via vCenter UI. These settings can be seen in Figure 4.7.

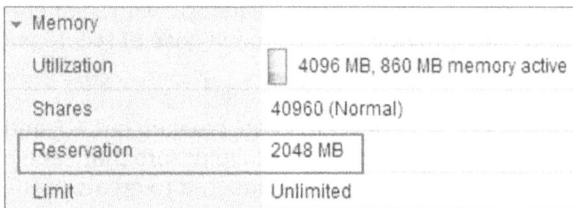

▾ Memory	
Utilization	🖩 4096 MB, 860 MB memory active
Shares	40960 (Normal)
Reservation	2048 MB
Limit	Unlimited

Figure 4.7 Default NSX Controller Settings

Note that similar to NSX Manager, NSX controllers are not in the data path. In case one controller goes down, it will not impact the traffic because the other controller VM in the cluster will assume responsibilities for the functions performed by the failed VM.

DLR Control VM

The DLR control VM is only needed if a dynamic routing protocol is required between the DLR and ESG. As a best practice, businesses are encouraged to use a dynamic routing protocol in these configurations. In small data centers, an organization might choose to deploy a DLR with the static routing. In that situation, a DLR control VM is not needed, saving some compute capacity.

An anti-affinity rule is created automatically if a DLR control VM is deployed with HA option. This is only possible if DRS is enabled on the cluster.

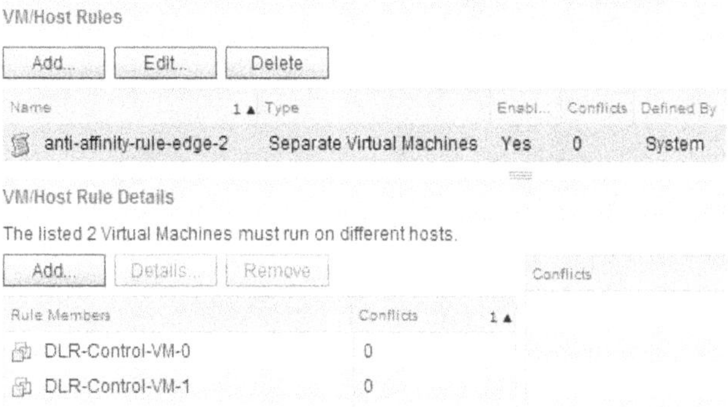

Figure 4.8 DLR Control VM Default Anti-Affinity Rules

These VMs must be deployed in active/standby HA mode for production deployment.

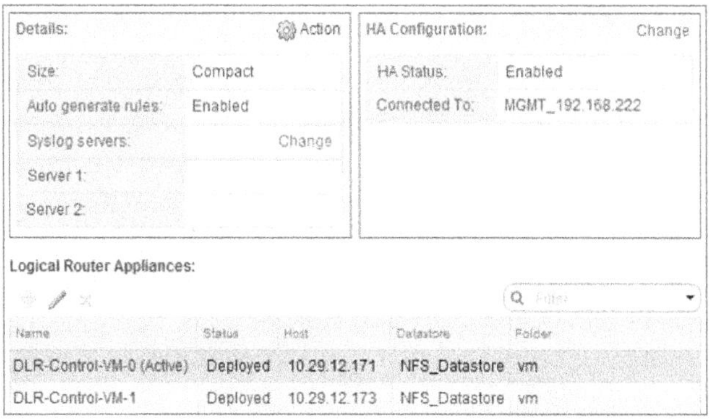

Figure 4.9 DLR Control VM Size and Active/Standby Status

Notice that these VMs are also locked-down VMs and CPU/memory modification is disabled.

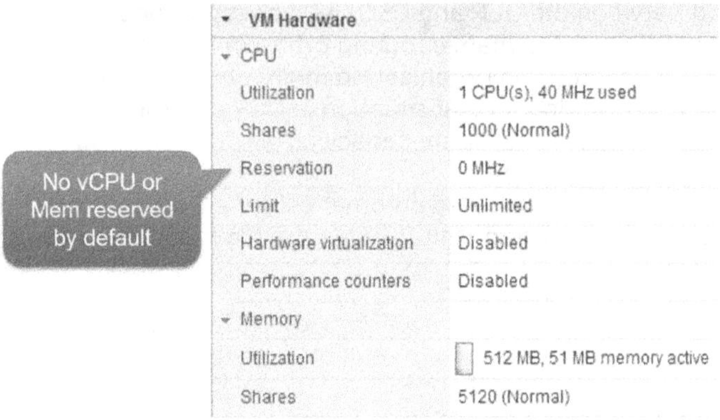

▼ VM Hardware	
▼ CPU	
Utilization	1 CPU(s), 40 MHz used
Shares	1000 (Normal)
Reservation	0 MHz
Limit	Unlimited
Hardware virtualization	Disabled
Performance counters	Disabled
▼ Memory	
Utilization	512 MB, 51 MB memory active
Shares	5120 (Normal)

No vCPU or Mem reserved by default

Figure 4.10 DLR Control VM Default Settings

Edge Services Gateway (ESG)

The ESG VM sits at the edge of the NSX virtual network and provides on-ramp/off-ramp connectivity between the virtual overlay and physical underlay. The ESG VM can be deployed into two different modes:

1. ESG with HA

2. ESG with ECMP

The ESG with HA option deploys two ESG VMs, one acting as primary with the second in standby mode. In small data centers, ESG with HA is the most deployed option because it allows organizations to enable stateful services such as DHCP, NAT, and VPN.

With ESGs deployed in ECMP mode, up to eight VMs can be active and participate in forwarding traffic in/out from the virtual network to the physical undelay. Since all ESG VMs are active with ECMP, the traffic may exit via one ESG VM and come back in through another. Because of this , it is not possible to run stateful services on ESGs in ECMP mode. Usually ESG with ECMP is deployed when more than 10G of north-south bandwidth is required.

The following decision tree is a simplified guide to help decide which ESG deployment mode is best based on customer requirements and growth forecasts.

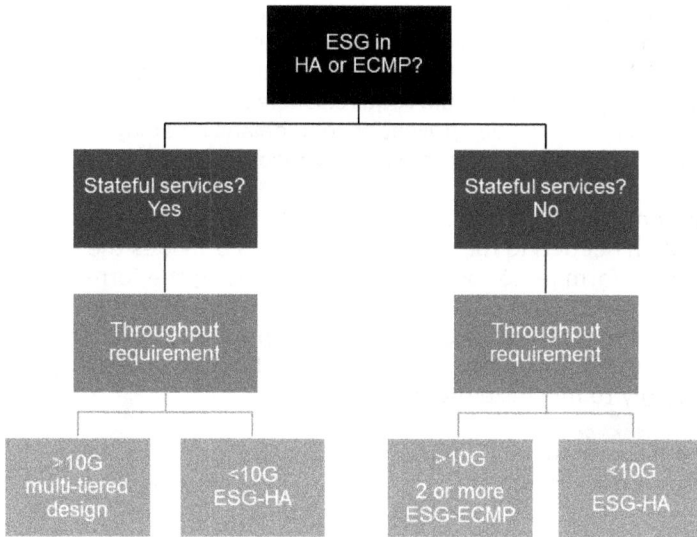

Figure 4.11 ESG mode selection decision tree

ESG Multi-Tiered Design

In multi-tier ESG design, an organization might deploy an ESG in Tier0 and Tier1 fashion. Tier0 ESGs are configured in ECMP mode for higher throughput at the boundary while tier1 ESGs are placed close to workload VMs and configured with HA to provide stateful services.

Although not common, the ESG multi-tiered design could be used in small data centers in scenarios where deployment demands higher throughput with stateful services.

ESG multi-tiered design is outside the scope of this book. Readers are encouraged to consult with the *VMware NSX for vSphere Network Virtualized Design Guide* for further details on this topic, located at:

https://communities.vmware.com/docs/DOC-27683

The aim of this book is to discuss the most popular designs for SMB. Other designs are also possible depending on the specific requirements and scale of the deployment.

General ESG Deployment Considerations for Small Data Centers

An ESG VM can be deployed in various form factors. For majority of small data center deployments, the large form factor is recommended. The ESG VM large form factor is sufficient for majority of the services that a small data center would require. One exception is a situation with a requirement to provide application load balancing at layer 7; in this case an extra-large form factor is highly recommended.

NSX automatically creates an ESG VM with the vCPU/memory reservation needed to run smoothly. NSX also provides the option to change the form factor of the ESG VM. Changing the form factor will redeploy the ESG VM but it will retain its existing configuration. Changing the form factor could be a disruptive event depending on the services being configured; this action should be planned accordingly to minimize the downtime.

Table 4.4 shows the ESG VM form factors and their associated resource requirements.

Table 4.4 ESG VM Form Factors

VM For Factor	vCPU	Memory	Disk Space	Suitable For
Compact	1	512 MB	~ 500 MB	Lab/PoC environment
Large	2	1 GB	~ 1024 MB	All services except L7 load balancer
Quad Large	4	1 GB	~ 1024 MB	
X-Large	6	8 GB	~ 2500 MB	Layer 7 load balancer

ESG Deployed in HA

When an ESG is deployed with the HA option, the anti-affinity rules are automatically created if the cluster is enabled with DRS. DRS will ensure the active/standby VMs are not running on the same ESXi host.

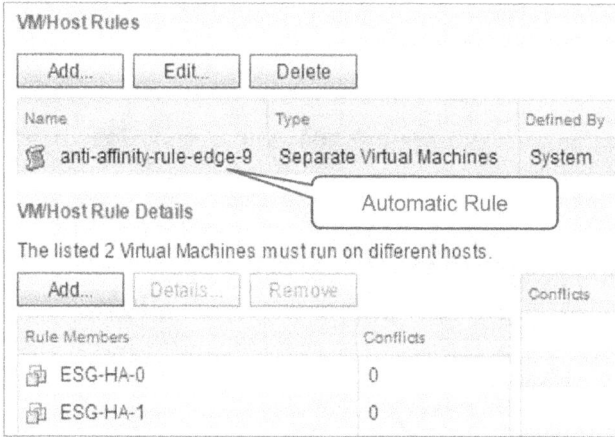

Figure 4.12 ESG deployed in HA with automatic anti-affinity rule creation

If DRS is not enabled on the cluster, then the business should manually make sure that active and standby VMs are not residing on the same ESXi host at any time. If the active and standby ESGs are co-resident on an ESXi host that loses connectivity or goes down, there may be a service disruption. Recommended design configuration with an ESG in HA is shown in Figure 4.12.

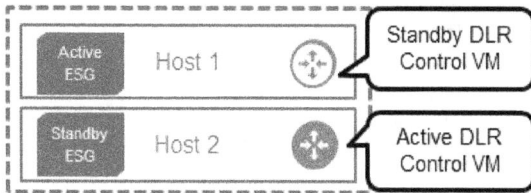

Figure 4.13 VM placement on ESXi host for ESG in HA with DRL Control VM

ESG Deployed in ECMP

With ESGs configured in EMCP mode, all ESG VMs are active. They will actively participate in forwarding traffic. If one ECMP ESG VM goes down, the traffic will still flow through other active ECMP ESG VMs. With the ESG in ECMP mode, make sure to disable ESG firewall and ESG HA options in the NSX UI. Enabling both ECMP and ESG HA options will create an extra, unnecessary standby VM per ESG VM. Another design consideration is to make sure that an ECMP-enabled ESG VM is not deployed on the same ESXi host with the active DLR control VM. Make sure to create anti-affinity rule so that the distribution and placement of VMs follows recommended guidelines.

Recommended design configuration is shown in Figure 4.13.

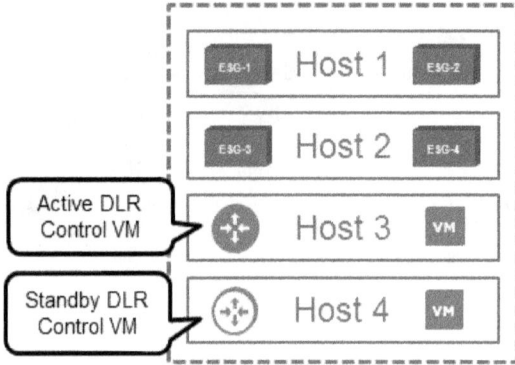

Figure 4.14 VM placement on ESXi host for ESG in ECMP with DLR Control VM

VDS Considerations

The VDS is the only virtual switch that is supported for use with NSX; vSphere Standard Switch (VSS) and other 3rd party vendor virtual switches are not supported. The VDS is included with vSphere Enterprise+; however, the VDS is included with the purchase of the NSX license regardless of the type of vSphere license.

In large data centers there could be multiple clusters and multiple VDSs. These may be based on various design factors such as a unique VDSs for management, payload, and Edge traffic. In a small data center, there is typically a single cluster running all the VMs for management, payload/workload, and Edge functionality. In order to keep the design simple in a small data center with single cluster, it is recommended to use a single VDS per vCenter. Multiple port groups can still be used in the single VDS design.

In addition to the port groups manually created by the administrator in VDS, VTEP port groups and VM kernel port(s) are automatically created based on the NIC teaming selection at the time of preparing for the vSphere cluster for NSX.

In large and medium data centers, the recommended VTEP vmknic teaming policy is route based on originating port (source-ID). This teaming policy will create multiple VTEP ports per host (if multiple physical NICs are present) in order to achieve multi-pathing. This option provides multiple VXLAN paths for traffic leaving the ESXi host.

The recommendation to select source-ID as vmknic teaming policy might not be the desirable choice for some SMB data centers. SMB data centers are more inclined towards simple VDS design for configuration and troubleshooting. SMB data centers can choose the NIC fail-over option as teaming policy which will create a single VTEP per ESXi host.

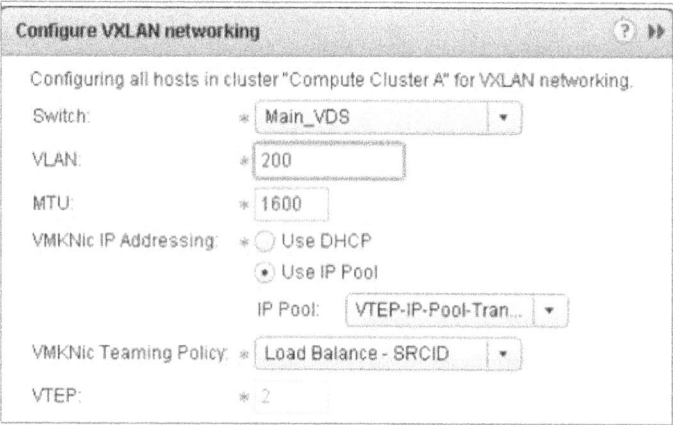

Figure 4.15 VXLAN networking options during NSX host prep operation

Chapter 4 - Key Takeaways

- NSX is supported on all vSphere license types.

- VDS is part of NSX license regardless of the type of vSphere license.

- NSX VM exclusion list must include vCenter /PSC VM unless fine grained rules are created in NSX to allow access to vCenter/PSC VMs.

- Create anti-affinity rules for NSX controllers with the "SHOULD" option.

- The majority of SMBs have deployed ESG with HA.

- vCenter should be the first VM in the VM boot-order list followed by NSX Manager and controllers respectively

NSX Operations: Monitoring and Troubleshooting

Regardless of the size of the data center deployment, day 2 operations and troubleshooting aspects are equally as important to consider as the initial design and planning. Operations teams are often not involved during the design phases, leading to decisions detrimental to implementation and ongoing operations.

Building software defined virtual networks requires all teams – including network, infrastructure, and security – to be involved during the phases of design, implementation, and operations. Some organizations create cloud admin teams which include representation from various data center teams. In SMB data centers there are usually only a few people working closely together who are responsible for all design, implementation, and day-to-day operations. In such environment, there is no requirement to create a formal cloud team.

The following section discusses the tools and strategies that can be adopted by IT directors and managers of small data centers.

Networking and Security Operations Requirement

The following are key areas where data center teams should focus efforts to enhance day 2 operational efficiency:

- Monitoring – continuous and proactive monitoring of all components including NSX, non-NSX, physical, and virtual

- Troubleshooting – standard procedures to troubleshoot and diagnose problems in the event of a failure or critical error

- Change and incident management

- Performance and capacity planning

Of these four areas, this book will focus on the monitoring and troubleshooting tools and techniques necessary for successful data center operation. Since change and incident management are more relevant to people and process, those topics are outside the scope of this book. Similarly, this book does not cover performance monitoring, characterization, or capacity planning in detail.

Monitoring and Troubleshooting Tools and Products

NSX provides a great deal of monitoring and visibility into data center operations through a variety of distinct toolsets. The first set of tools is provided by NSX Manager itself; these are the tools that the NSX platform brings to the table without additional investment.

NSX Native Tools

These native NSX tools are of interest in the small data center because there is no additional cost to acquire them; they come with any NSX licensing scheme. These native tools include:

- NSX Manager GUI

- NSX Central CLI

- NSX RESTful API

- Standard protocols and methods

  - Port mirroring

  - IPFIX (Internet Protocol Flow Information Export)

- SNMP (Simple Network Management Protocol)

- Syslog

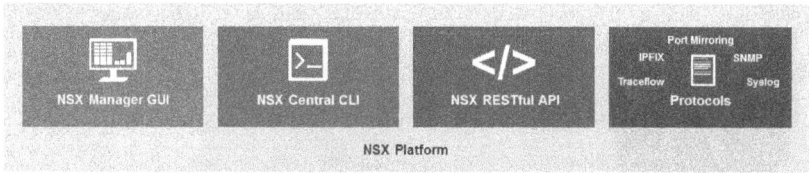

| NSX Manager GUI | NSX Central CLI | NSX RESTful API | Port Mirroring / IPFIX / Traceflow / SNMP / Syslog Protocols |

NSX Platform

Figure 5.1 NSX Native Tools and Protocols

VMware Products

In addition to natively available tools and protocols, there are other VMware offerings that tightly integrate into NSX and help with monitoring, operations, and visibility. The most popular and widely deployed products include:

- VMware vRealize® Log Insight™ (vRLI)

- VMware vRealize® Operations™ (vROps)

- VMware vRealize® Network Insight™ (vRNI)

vRealize Log Insight is a standards-based syslog server. It provides both standardized out-of-the-box dashboards as well as the capability to create and export custom views. vRealize Log Insight is by far the most utilized and deployed product in small data centers as it comes free with the NSX license.

vROps and vRNI provide enhanced visibility, monitoring, and alerting capabilities for overlay and underlay networks and are available as separately licensed products.

3rd Party Products

Besides NSX native tools and VMware ecosystem products, there are 3rd party certified partner products that help customers monitor and troubleshoot NSX and data center operations. Vendors like Splunk and Gigamon offer plugins and hooks that connect to NSX and vCenter to pull the relevant logs and data from these systems. These offerings provide further insight into the network and security environments.

Small data center operators and architects must evaluate the requirement and benefits of using these products based on their current environments and growth plans. These 3rd party vendor monitoring and troubleshooting products are outside the scope of this book. Readers interested in knowing more about these products can refer to the following URL:

https://www.vmware.com/products/nsx/technology-partners.html

Troubleshooting Methodology

When troubleshooting any environment, it is important to break down and isolate the problem area. Since NSX uses a network overlay to achieve network virtualization, it is important to understand the relationship between the NSX virtualized environment and the physical underlay. Breaking down troubleshooting into small stages is a proven approach in the industry and has been adopted by many IT operations teams.

These stages could vary based on type of business and scope of data center operations. In a very large data center, the initial stages may be outsourced to a managed service provider for local or remote support. In a small data center there is an assumption that teams are local, at least at the beginning of the project. Hence, multiple stages can be handled by the same team and people.

The stages listed below provide a high level overview and troubleshooting methodology that can easily be adopted in a small data center.

Level 0 Troubleshooting Stage

The level 0 stage is essentially the documentation stage. Every organization must have good documentation explaining the design, configuration, and steady state packet flow. Businesses must have the following information before beginning troubleshooting:

- Understanding and documentation of physical + virtual topology

- Understanding and documentation of physical + virtual packet flow in steady state

Level 1/Level 2 Troubleshooting Stages

At this stage, the troubleshooting is done using high level tools and technique to narrow and isolate the problem before going into advance debugging and troubleshooting mode. This stage is also useful to identify whether the incident is related to the physical underlay or NSX overlay.

Typically, logs are collected and reviewed at this stage to find out the timing of the incident. Logs can be collected from various NSX and non-NSX components. These logs provide insight between the historic and current information to triage the incident. Logs can be correlated between devices to help isolate timing and event progression throughout the environment.

Following is a list of recommended tools that could be used during this troubleshooting stage:

- NSX monitoring dashboards – NSX and cluster health indication

- NSX Traceflow – visual virtual network packet trace

- NSX flow monitoring – quick packet flow

- NSX endpoint monitoring – monitor OS processes running inside a VM

- vRealize Log Insight – presence of this or another syslog server is a must

- vRealize Network Insight – optional in a small data center

Level 3 Troubleshooting Stage

If the incident cannot be resolved at level 1 or level 2, the next step is to employ advance tools and techniques provided by NSX. Following are the very powerful tools that are available within NSX:

- NSX Central CLI

- Packet captures

The NSX Central CLI reduces troubleshooting time for distributed network functions. Commands are run from the NSX Manager command line – accessible via ssh – and can retrieve information from the NSX controllers, ESXi hosts, and NSX Manager itself. This allows quick access and comparison of information from multiple sources. The Central CLI provides information about logical switches, logical routers, distributed firewall, and Edges.

Packet captures – or tcpdumps – are extremely powerful tools that are provided with NSX and vSphere. The ESXi CLI can execute "pktcap-uw" and capture a complete dump of traffic entering or leaving ESXi hosts at various stages within the packet flow.

The NSX Edge VM provides a CLI to assist with troubleshooting. That Edge CLI provides a "debug packet" command that can be used to take a TCP dump of the traffic leaving and/or entering the NSX Edge

VM, effectively providing the same functionality as the "pktcap-uw" command from the ESXi host.

The above list is not exhaustive; many additional tools and techniques are provided by 3rd parties.

For more details on troubleshooting NSX topic, refer to following URL:

https://docs.vmware.com/en/VMware-NSX-for-vSphere

Performance and Capacity Planning

It is important to understand that, like any other networking design, there are consolidation and oversubscription ratios that should be monitored on the ongoing basis. In some cases, small data center implementations start with limited resource allocation which are increased to accommodate new requirements as workloads grow. This under-allocation of resources could jeopardize NSX operations, resulting in application outages or poor performance.

Proper capacity planning and monitoring can avoid such situations, minimizing the risks that are associated with lack of monitoring growth relative to the original scope and design of the small data center.

Data centers are designed based on specific workload requirements; it is imperative to add resources as deployments grow. There are various options and tools available for monitoring and capacity planning, some listed here:

- Native vSphere/ESXi/vCenter tools

    - vSphere alerts

    - ESXi CLI commands (e.g., esxtop)

    - Customized PowerCLI scripts

- Native NSX Tools

    - NSX component UI and alerts

    - Customer dashboard with NSX REST API

- VMware vRealize Operations

- vRealize Network Insight

In a small data center, the budget is often a constraint. In these situations, native vSphere and NSX tools can provide basic, but adequate, performance and capacity planning.

Chapter 5 - Key Takeaways

- Regardless of the size of the data center, day 2 operations play an important role in the overall environment and application availability.

- A well-documented troubleshooting methodology should be employed based on specific data center needs and human resources.

- NSX provides native tools and industry standard protocols in order to monitor and troubleshoot data centers.

- NSX integrates with VMware products such as vRealize Log Insight, vRealize Operations, and vRealize Network Insight to provide out-of-the-box dashboards and plugins for visibility, monitoring, and troubleshooting both underlay and overlay networks.

- Continuous monitoring of resource utilization and related adjustment based on new or existing application demand is necessary for customer experience and business success.

Growing NSX Deployments

NSX enables network virtualization by defining networking and security services in software. NSX is highly flexible and versatile, providing options to grow without re-architecting the physical underlay. Growing and expanding an NSX deployment is possible in multiple directions, broadly categorized into two sections:

- Growth within the small data center

- Growth beyond small data center

Growing NSX within Small Data Center

Figure 6.1 provides some of the growth possibilities and options used by existing NSX customers; many deployments started with a small footprint and grew in one of these dimensions.

Figure 6.1 NSX growth options and possibilities

The following sections will discuss each area one by one and expand on possibilities to grow NSX within a small DC design.

Growing NSX with Additional Compute Capacity

A popular NSX growth dimension is the addition of compute capacity. After deploying NSX in a small data center, many businesses realize the benefits and quickly add additional ESXi hosts to their deployment.

This can be done either by adding hosts to the existing cluster or by creating a new compute cluster. At a very high level, adding one or two additional ESXi hosts to an existing cluster does not require significant redesign. It can be done relatively easily and without disruption.

The decision between adding hosts or creating a new cluster should spark discussion within an organization. The discussion should examine the type and size of workloads, throughput requirements, total number of hosts being added, and future growth potential. These options are discussed in detail in the later part of this book under "Small to Medium" and "Small to Large" scale data center design.

Growing NSX with Additional Throughput

Depending on the demands of applications, the NSX small data center design can be easily modified to provide additional bandwidth. Depending on the underlay resource, application requirements, and stateful service needs, the additional throughput can be increased by adding more ESG VMs.

If a small data center has adopted the ESG HA mode with stateful services, convert the NSX Edge design into an ECMP based design to add more bandwidth and a multi-layer ESG design. If a small data center has deployed the ESG HA mode without the stateful services, convert the NSX Edge HA into a single tier ECMP design by making configuration changes through the NSX Manager UI.

In both options, as a best practice, make configuration changes during the maintenance window and plan for an outage during changeover and testing. Proper planning and monitoring is necessary to make sure adequate throughput is available to applications/VMs – otherwise applications/VMs might suffer performance issues.

Growing NSX with Additional Services

NSX provides networking and security services either in distributed fashion or centralized via an ESG VM.

The following is a list of distributed services offered directly in the ESXi hypervisor kernel:

• Logical switch (VXLAN)

• Logical router (DLR)

• Distributed firewall (DFW)

Edge services offered directly via the ESG VM include:

• Firewall

• DHCP

• NAT

- Switching (VXLAN and VLAN)

- Routing

- Load balancer

- L2 VPN

- L3 IPSec VPN

- SSL VPN

An organization may start with one or two NSX services and, with proper planning, grow non-disruptively while adding new services to the existing deployment without re-architecting the overall design.

For instance, an organization can start with micro-segmentation as a use case, then move to the NSX full stack deployment model with VXLAN and dynamic routing. This can be accomplished by simply making configuration changes via the NSX Manager UI. An organization planning this change should discuss it with their physical network/security teams to make sure the correct MTU is configured and firewall ports are opened in the physical underlay. Dynamic routing should also be configured to the physical routing infrastructure for route peering with the NSX ESG.

Growing NSX with Additional Sites

Businesses can also take advantage of NSX cross-vCenter capabilities to add more sites for resource pooling, business continuity, and disaster recovery. There are businesses that initially deployed NSX in a small data center and then moved to harness the potential advantages of a cross-vCenter solution. These businesses enabled NSX to support a multi-site solution.

Adding additional sites with cross-vCenter and SRM-based solutions requires a deeper discussion and additional design consideration that are outside of the scope of this book. Readers are encouraged to review the NSX Cross-VC Design Guide provided at the following link:

https://communities.vmware.com/docs/DOC-32552

Also refer to VMware NSX documentation for requirements and implementation guidelines:

https://www.vmware.com/support/pubs/nsx_pubs.html

These guides are not specifically written for the small data center, but the concepts presented therein can easily be adopted for small data center designs and deployments.

Growing NSX with Data Center or Workload Migration

Workload migration from physical to virtual (P2V) or from virtual to virtual are compelling NSX use cases. NSX has helped businesses migrate specific workloads or their entire payload infrastructure to a virtual environment. Specific to migration, NSX provides capabilities such as:

- NSX cross-vCenter, coupled with vSphere L3 vMotion capabilities

- All-software VXLAN based L2 extension across sites (kernel-level distributed service)

- L2VPN to extend VLAN to VXLAN (ESG based)

- NSX L2 software bridging (DLR control VM based)

Two themes have emerged from the wide spectrum of organizations growing their NSX deployments by migrating workloads:

- Workload migration from a legacy physical environment to NSX-based virtual network

- Workload migration from existing vSphere VLAN -backed port groups to NSX-based virtual network

More details on this topic is available at NSX Brown Field Deployment Guide:
https://communities.vmware.com/docs/DOC-29556

Readers are highly encouraged to refer to this document as the concepts presented in the *NSX Brown Field Deployment Guide* can also be applicable to small data center scenarios.

Growing NSX Beyond Small Data Center

Previous sections discussed growth options and scenarios related to small data centers, typically with a single vSphere cluster. What if the data center size is increasing, more hosts are being added, or bandwidth requirement are growing such that it requires the addition of a large number of ESXi hosts?

If the data center footprint is increasing drastically or is forecast to in the future, it may indicate that data center re-engineering is required to support the intended increase. From an NSX perspective, increasing from a small data center to a medium or large data center is relatively straightforward and does not require significant changes.

vSphere cluster layout, availability, service selection, and placement of infrastructure VMs are important factors to considering when growing the data center size. Special focus is given to these areas in the following sections.

It is not possible to cover every possible growth scenario and use case in this book. The aim of this section is to cover the most popular situations and use cases faced by customers looking to grow NSX beyond the small data center model.

Growing NSX from Small to Medium Data Center

It is a common scenario to grow NSX deployment from a small to medium data center design; organizations are often eager to realize the advantages and benefits that NSX brings with its operational simplicity and ease of use.

Businesses should recognize and assess the indicators to grow from small to medium data center design. Following are some of the commonly asked questions on this topic by business owners and decision makers:

- What are the factors and indicators to gauge that it is time to grow NSX deployment?

- What could be the challenges associated with this growth?

- Does this growth require re-architecting the current data center and/or NSX design?

- Does it require creating separate vSphere clusters?

While there is no one size fits all approach, an organization should consider following factors to help with this decision:

- Network utilization

- Features and use cases deployed

- Business requirements and application growth pattern

- Separation of management and edge components from payload/application VMs

- Operational simplicity and clear demarcation

- Predictive traffic pattern

As a rule of thumb, organizations can start thinking about growing from a small data center design to a medium data center design if following changes are observed or planned:

- If an organization is planning to add somewhere between 10 to 100 additional ESXi hosts, running hundreds of VMs, to its existing footprint.

- If an organization's north-south bandwidth requirement (i.e., between virtual network and physical underlay) is less than but approaching 10Gig.

- If the business requirement is to create dedicated payload cluster(s).

- If an organization wants to restrict the ESG to physical TOR/ firewall VLAN connectivity to a certain cluster.

The following section discusses the design principles and considerations for medium data center cluster design.

Medium Data Center Cluster Design

Figure 6.2 depicts the medium data center cluster layout. In a medium data center cluster design, organizations can add more ESXi hosts and create a separate payload cluster for application/workloads. The management and Edge components may stay in the existing collapsed cluster, shown in the figure as "Collapsed Management & Edge Clusters"

Figure 6.2 further shows a possible distribution of management and Edge VMs to protect against a single point of failure. Similar to small data center design, since the bandwidth requirement is not greater than 10G, the organization can continue to use the NSX Edges with HA that would allow it to run stateful services.

Figure 6.2 Medium Size DC Cluster Layout and infra. VM placement

It is important to highlight that this book is not covering all the

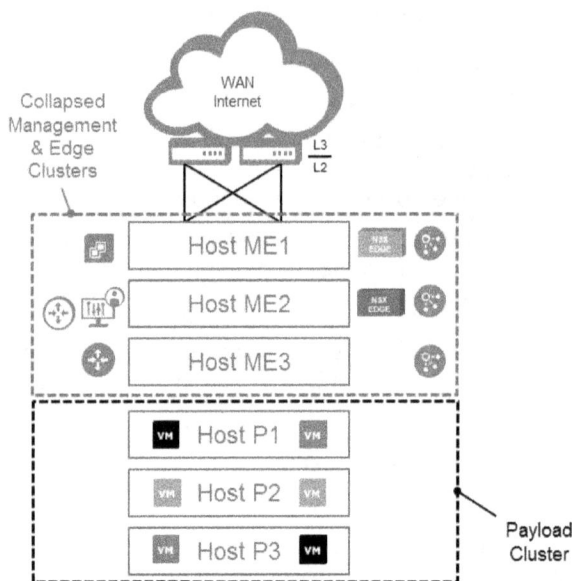

Collapsed
Management
& Edge
Clusters

WAN
Internet

L3
L2

Host ME1

Host ME2

Host ME3

Host P1

Host P2

Host P3

Payload
Cluster

implementation level details required to grow NSX to a medium data center. The intent is to list the important possibilities and approaches to grow NSX accordingly. Additional discussion and requirement evaluation is necessary to implement this design. Businesses should reach out to their VMware account team or preferred integration partner when they are ready to have these conversations around growth. The VMware account team can provide recommendations and include the NSBU Customer Success Solutions Architect team in these discussions to help organizations achieve their desired goals.

Growing NSX from Small to Large Data Center

Many organizations want to know when is the perfect time to increase the size of their data center footprint. Questions on this topic are listed below, and do not differ from the discussion in the "Growing NSX from Small to Medium Data Center" section.

- When to move from small data center and single cluster design to a large size data center?

- Does this growth require re-architecting current data center and/ or NSX design?

As discussed in the previous section, the answers to these questions depend on many factors:

- Current network utilization

- Features and use cases deployed

- Business and application growth pattern

- Acquisition/mergers

- Operational model

As a general rule of thumb, an organization can start thinking about growing from small NSX design to large data center design if following changes are observed or these events are being planned:

- If an organization is planning to add more than 100 ESXi hosts, running thousands of VMs, to its existing data center footprint.

- If its north-south (e.g., between virtual network and physical underlay) bandwidth requirement is greater than 10Gig.

The following section discusses the design principles and considerations for NSX large data center cluster design.

Large Data Center Cluster Design

In a large data center design, it is recommended to separate management, Edge and payload functions. This separation helps organizations easily add compute capacity or ESXi hosts to existing or new clusters based on business need.

The logical separation and grouping of the ESXi hosts to provide specific functions such as payload, management, and Edge services provides the following advantages to the data center architect:

- Flexibility of expanding and contracting resources for specific functions

- Ability to isolate and develop span of control over various control plane components, such as control VM, Edge VM deployment, and other technology integration

- Managing the lifecycle of certain resources for specific functions (e.g., socket/core count, memory, NIC, upgrade, migration).

- High availability based on functional needs

- Isolate automation to specific resources (e.g., app-tier, security tags, policies, load balancer)

Figure 6.3 shows separate management, Edge, and payload clusters hosting respective workloads/VMs in a large data center design. It also shows the possible distribution of the various management and Edge VMs on separate hosts within the cluster to avoid a single point of failure.

Management Cluster

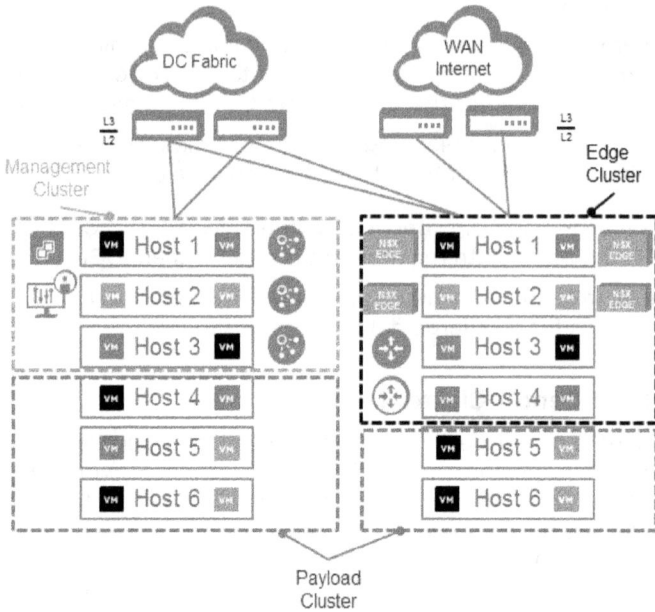

Payload
Cluster

The management cluster hosts the management components, including vCenter Server, NSX Manager, NSX controller, and Cloud Management Systems (CMS). Compute and memory requirements for hosts and resources are typically pre-identified based on the required scale and minimum supported configurations. Capacity and resource requirements within the cluster are fairly consistent. The availability of components is important, and the hosts in the management cluster can improve resiliency by enabling LACP.

Unlike small and medium design, the ESXi hosts in management cluster do not require VXLAN provisioning.

Edge Cluster

The Edge cluster would typically host the ESG and DLR Control VM. ESGs provide critical interaction between the physical infrastructure and overlay network. The Edge cluster has following characteristics and functions:

- Provide on-ramp and off-ramp connectivity to physical networks (i.e., north-south L3 routing on NSX Edge virtual appliances).

- NSX controllers should be hosted in an Edge cluster when a dedicated vCenter is used to manage the payload and Edge resources.

- Edge VMs have specific connectivity needs and characteristics:

  - Edge VM forwarding is CPU-centric with consistent memory requirements

  - Edge resources require external connectivity to physical network devices, possibly constraining physical location placement to minimize VLAN spread

  - Recommended teaming option for VDS on Edge hosts is "route based on source ID"

- ESG VMs are typically deployed with ECMP to achieve greater bandwidth requirements in a large data center design. These ESG VMs have anti-affinity requirements to prevent all of them from running on the same host at steady state while also ensuring ECMP Edge and active DLR control VMs are never running on the same ESXi host

  - Figure 6.3 shows a possible placement of ECMP ESG VMs, where two ECMP ESG VMs are running per ESXi host. It is assumed that this client has NSX bridging, so Host3 and Host4 are reserved to run the DLR Control VM.

  - If bridging is not a requirement, an alternate approach would place three ESXi hosts in the Edge cluster – one extra in case a host with ECMP Edges goes down – and place the DLR control VM in the payload cluster.

Table 6.1 shows the recommended number of hosts required in the Edge cluster based on use case with DLR control VM placement in the Edge vs. payload cluster.

Table 6.1 Recommended number of hosts with DRL VM placement options

Use-Case	Hosts in Edge Cluster	Comments
ECMP with NSX Bridging	4	2 hosts for ECMP ESG VMs and remaining two hosts to run DLR Control VM for dynamic routing and DLR Control VMs for bridging instance in edge cluster Anti-Affinity rule to make sure ECMP ESG VMs "should" not be allowed to run on the hosts where Active DLR Control VM resides
ECMP without NSX Bridging	3	2 hosts for ECMP ESG VMs and one spare host (This requirement could reduce to 2 hosts if N-S bandwidth requirement is less than 20GiG) DLR Control VM running in the payload cluster

Payload Cluster

The payload cluster is part of the infrastructure where workloads are provisioned and application VM connectivity is enabled via logical networks. In large data center designs it is important to consider following aspects:

- Total number of clusters and the number of ESXi hosts per cluster. These numbers should adhere to VMware recommended sizing guidelines.

- Rack-based cluster vs. multi-rack/horizontal striped cluster

- Workload centric resource allocation

NSX Growth Scenario

There are many different NSX growth scenarios; an entire book could be dedicated to covering use cases and deployment options. In this instance, a single scenario is covered to reinforce the concepts that have been discussed throughout this book. Organizations and architects are encouraged to consult their VMware account teams or partners to discuss the details and other scenarios.

Customer Case Study

The scenario mentioned here is from a real customer. To protect the identity of this customer, it is referred to here as ABCH (Americas Best Clinical Healthcare).

ABCH is a large healthcare provider with branches and operations in more than 12 states in the USA. ABCH has four large data centers in different states. ABCH started with a micro-segmentation use case with small data center footprint to begin its NSX network virtualization journey. ABCH deployed NSX using the security focused deployment model as shown in Figure 6.4. It was a non-disruptive deployment that enabled micro-segmentation while using VLAN backed port groups.

Figure 6.4 ABCH Security Focused Deployment

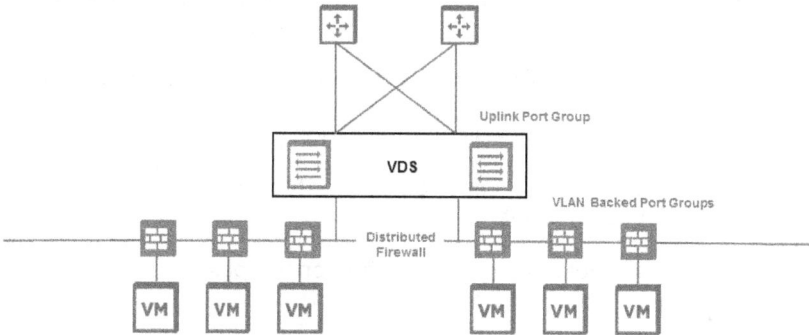

Security focused deployment with micro-segmentation helped ABCH clear their PCI compliance audit by maintaining clear boundaries between PCI and non-PCI workloads without requiring physical air-gaps between DMZ environments. The operational simplicity was a huge advantage for the ABCH IT operations team, as security policies were automatically added to or removed from the payload without manual intervention. The success of NSX prompted ABCH to take advantage of NSX's full potential with the NSX full stack deployment model.

ABCH decided to grow NSX to a full stack deployment model to take advantage of following features:

• Logical switching (VXLAN) : decoupling from the physical constructs and the limited number of VLAN IDs

• Logical routing (DLR) : optimized routing within the host and inside the virtual network to avoid hair-pinning to physical firewall

Growing an NSX security focused deployment into full stack deployment requires interaction with physical infrastructure. Designing for virtual infrastructure on the physical underlay requires detailed discussions with the network and security teams. The various ABCH teams held joint design discussions and workshops with VMware NSBU customer success solutions architects to make sure all the requirements were gathered. The constraints were captured and the risks were discussed. All of this information was documented and incorporated into their mitigation plan.

Summary of Design Discussion

This book cannot cover the entire architecture or all of the documented design decisions, but from a high level, expanding from security focused deployment mode to full stack requires the following:

- The MTU for the physical VLAN that will carry the VXLAN based overlay traffic (i.e., VTEP to VTEP traffic) should be at least 1600. The recommendation is to set the MTU to jumbo.

- IP addresses/subnet space should be planned and carved out for:

  - VTEP to VTEP VXLAN communication

    - This is a physical and routed VLAN as mentioned in the first bullet point.

    - This should be large enough to encompass current and future host requirements

  - VXLAN transit logical switch

    - This is a small subnet space for IPs to be assigned for communication between an Edge Services Gateway (ESG) and Distributed Logical Router (DLR)

    - This is a floating IP range that is routable but does not have a VLAN associated to it in the physical underlay

  - Workload VMs that will be connected to logical switches

    - This will depend on the number of VMs attached

    - The recommendation is to use a subnet no larger than /22 (~1K IPs) for a logical switch

    - This is a floating and routable subnet space without an associated VLAN in the physical underlay

- VTEP IP addresses will be configured on ESXi hosts as part of the VXLAN networking configuration step. The number of VTEP IP addresses per host depends on the NIC teaming type used on the VDS.

- If dynamic routing is needed in the virtual network, it would require deployment of a DLR control VM in HA mode.

- NSX ESG VMs will be deployed in the either ECMP or HA mode depending on business requirement. Addition of an ESG into the design will require changes on the physical network.

- The default gateway for the VMs attached to logical switches will point to the NSX DLR.

  - This does not require changing IP addresses on workload VMs

    - This does require changes on the physical network. This transition should occur when the default gateway is moved to the virtual infrastructure's DLR.

- As part of the expansion, 3rd party services can be added for features like guest introspection and network introspection.

A high level architectural diagram is shown in Figure 6.5. The workload VMs are connected to logical switches, referred to as VXLAN backed port-group in the diagram. The DLR is introduced and will route the traffic between workloads subnets in the logical space. The DLR will eliminate the traffic hair-pinning to the physical firewall/router, saving bandwidth and optimizing traffic flows.

Figure 6.5 ABCH Full Stack Deployment

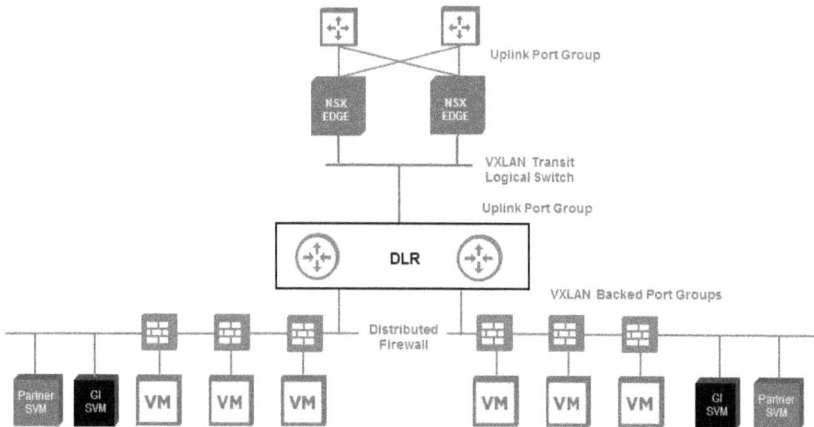

Chapter 6 - Key Takeaways

- NSX is highly flexible and can grow in any direction.

- Organizations can start from a use case based on business needs, and can grow into other use cases as needed without re-architecting their entire data center design.

- NSX data center growth could be:

  - Within the small data center by adding more features

  - Growing a small data center to a medium or large data center architecture

Chapter 7

Conclusion

The data center landscape is changing and at important crossroads. Growth in networking and security spend is vastly outpacing growth in overall IT spend. The growth and change is so rapid that traditional networking vendors are unable to keep pace. VMware is the game changer in the market, using its revolutionary network and security virtualization – powered by NSX – to re-define data centers entirely in software. SDDC is the only way for organizations to keep up with ever-increasing application demands, providing the flexibility and agility that they deserve. This demand is not limited to large organizations and operations, but also applicable to small and medium businesses and their data center requirements.

VMware is at the forefront of addressing the needs and requirements of these SMB customers. With its sophisticated, powerful, and flexible architecture, NSX offers the same features and power to small data centers that it has delivered to large enterprises and service provides across all verticals. The combination of NSX and vSphere provides a wide range of options and features that will ensure customer deployments are successful and that the full benefit of network virtualization technology and cloud enablement can be realized.

References

List of relevant references to design, deploy and manage NSX.

Planning and Design

NSX for vSphere Small Data Center Design Guide
https://communities.vmware.com/docs/DOC-33244

NSX for vSphere Network Virtualization Design Guide
https://communities.vmware.com/docs/DOC-27683

NSX for vSphere Cross-VC Design Guide
https://communities.vmware.com/docs/DOC-32552

NSX for vSphere Brownfield Design Guide
https://communities.vmware.com/docs/DOC-29556

Software Defined Data Center (SDDC) VMware Validated
Design Guide (VVD)
http://www.vmware.com/solutions/software-defined-datacenter/
validated-designs.html

Implementation

NSX for vSphere Installation Guides
https://docs.vmware.com/en/VMware-NSX-for-vSphere

VMworld 2017 - VMware NSX in Small Data Centers for Small
and Medium Businesses
https://youtu.be/a6SssLRIRNo

Operate and Manage

NSX for vSphere Operations Guide
https://communities.vmware.com/docs/DOC-30079

Index

The growth in public and private clouds spend is vastly outpacing the growth in overall IT spend. The change is so fast that traditional networking and security vendors are unable to keep pace. IT is looking at ways to keep up with the elastic demand and expectations from applications and the users in the world of Clouds. This trend is not only seen in large organizations, but can also be observed in small and medium businesses.

With network and security virtualization, VMware NSX is the game changer that can re-define data centers and build and run private clouds. VMware NSX is also the integration point between private and public cloud with offerings such as VMC (VMware Cloud) on AWS. VMware NSX, with its sophisticated, powerful yet flexible architecture, gives the same features and power to small and medium businesses as it has given it to large enterprises.

This book will help not only SMB but also large organizations to adopt this technology because it is seen that often large enterprises started their data center transformation journey with a small footprint. After realizing the huge impact and benefits of NSX, these large enterprises grew from small to medium or even large footprint in a short period.

The aim of this book is to give readers, architects, and engineers the necessary tools and techniques that they need to transform their legacy data center architecture to software defined private cloud based architecture. This book discusses a recipe of success, a well-orchestrated path to success, as well as a step by step approach to implement network and security virtualization that is proven and adopted by many in the industry.

vmware' PRESS

ISBN-13: 978-0-9986104-4-3
ISBN-10: 0-9986104-4-5

Cover design: VMware
Cover photo: 4X-image / iStock

ISBN 9780998610443

51299 >

www.vmware.com/go/run-nsx

9 780998 610443

$12.99